AN INTRODUCTION TO CHEMOINFORMATICS

AN INTRODUCTION TO CHEMOINFORMATICS

Revised Edition

by

ANDREW R. LEACH

GlaxoSmithKline Research and Development, Stevenage, UK

and

VALERIE J. GILLET

University of Sheffield, UK

 Springer

A C.I.P. Catalogue record for this book is available from the Library of Congress.

ISBN 978-1-4020-6290-2 (PB)
ISBN 978-1-4020-6291-9 (e-book)

Published by Springer,
P.O. Box 17, 3300 AA Dordrecht, The Netherlands.

www.springer.com

Printed on acid-free paper

CONTENTS

PREFACE

Chemoinformatics is concerned with the application of computational methods to tackle chemical problems, with particular emphasis on the manipulation of chemical structural information. The term was introduced in the late 1990s and is so new that there is not even any universal agreement on the correct spelling (cheminformatics is probably equally popular, and chemical informatics and even chemiinformatics have also been reported). Several attempts have been made to define chemoinformatics; among the more widely quoted are the following:

> The mixing of information resources to transform data into information, and information into knowledge, for the intended purpose of making better decisions faster in the arena of drug lead identification and optimisation. [Brown 1998]

> Chem(o)informatics is a generic term that encompasses the design, creation, organisation, management, retrieval, analysis, dissemination, visualisation and use of chemical information. [Paris 2000]

Many of the techniques used in chemoinformatics are in fact rather well established, being the result of years if not decades of research in academic, government and industrial laboratories. Indeed, it has been suggested that chemoinformatics is simply a new name for an old problem [Hann and Green 1999]. Whilst some of the current interest in chemoinformatics can be ascribed to the natural enthusiasm for things new, the main reason for its emergence can be traced to the need to deal with the vast quantities of data being generated by new approaches to drug discovery such as high-throughput screening and combinatorial chemistry [Russo 2002]. Concomitant increases in computer power, particularly for desktop machines, have provided the resources to deal with this deluge. Many other aspects of drug discovery also make use of chemoinformatics techniques, from the design of new synthetic routes by searching databases of known reactions through the construction of computational models such as Quantitative Structure–Activity Relationships that relate observed biological activity to chemical structure to the use of molecular docking programs to predict the three-dimensional structures of protein–ligand complexes in order to select a set of compounds for screening.

One characteristic of chemoinformatics is that the methods must generally be applicable to large numbers of molecules; this has been one of the principles that we have used when deciding what to include in this book. Our emphasis is on the computer manipulation of two- and three-dimensional chemical structures. We therefore include discussions of methods such as pharmacophore mapping and protein–ligand docking, which are perhaps more usually considered as "computational chemistry" rather than chemoinformatics, but which are now applied to large databases of molecules.

We start in Chapters 1 and 2 by considering the ways in which chemical structures can be represented in computer programs and databases. We then describe the calculation of molecular descriptors in Chapter 3; these form the basis for the construction of mathematical models such as Quantitative Structure–Activity Relationships described in Chapter 4. Chapters 5 and 6 deal with molecular similarity and similarity searching, and with methods to calculate molecular "diversity" and for selecting diverse sets of compounds. Chapter 7 follows with a discussion of methods for the analysis of data sets obtained from high-throughput screening. In Chapter 8 we consider approaches to virtual screening, including high-throughput docking and the use of *in silico* ADMET models. Finally, in Chapter 9 we describe methods for the design of combinatorial libraries. It is worth noting that certain methods (e.g. genetic algorithms and conformational analysis) are used in several areas; later sections may refer back as appropriate.

This book has had a rather long gestation period. One of us (ARL) had previously written a textbook on molecular modelling and computational chemistry [Leach 2001] and felt that relevant parts of the material would form a useful volume in their own right. Meanwhile, VJG was heavily involved in establishing one of the first academic courses in chemoinformatics [Schofield et al. 2001], was in need of a textbook and, in the absence of anything suitable, was planning to write her own. This book is the result of deciding to combine our efforts. Our target audience comprises graduate students, final-year undergraduates and professional scientists who wish to learn more about the subject. We have not assumed any prior knowledge other than some background knowledge of chemistry and some basic mathematical concepts.

We have tried to ensure that adequate coverage is given to the more widely used and available methods. However, it is not possible in a volume such as this to provide a definitive historical account or a comprehensive review of the field. The references and suggestions for further reading will hopefully provide good starting points for the inquisitive reader. We would be delighted to receive any constructive comments and feedback.

INTRODUCTION TO THE PAPERBACK EDITION

One indication of a maturing subject area is the publication of books and reviews intended for a general audience. The publication of the first (hardback) edition of this book together with others [Gasteiger and Engel 2003; Bajorath 2004; Oprea 2005] suggests that chemoinformatics has now reached this milestone. The subsequent four years have seen further advancements, both with regard to methodological developments and the practical application of the techniques in a number of fields. Publication of this new paperback edition has provided us with an opportunity to revise and update the text, which we hope will remain relevant not only to newcomers to the field but also to existing practitioners.

ACKNOWLEDGEMENTS

We would like to thank Peter Dolton, Drake Eggleston, Paul Gleeson, Darren Green, Gavin Harper, Christine Leach, Iain McLay, Stephen Pickett and Peter Willett who have read and commented on the manuscript. Particular thanks go to Peter Willett, who not only read the entire text but whose advice, expertise and encyclopaedic knowledge have been of immense value throughout the entire project; and to Drake Eggleston, who also read through the entire text and made some perceptive comments and suggestions. Naturally, any errors that remain are our responsibility – please do let us know if you find any.

We have been ably assisted by our team at Kluwer Academic Publishers (subsequently Springer): Peter Butler, Melissa Ramondetta, Tanja van Gaans and Mimi Breed and their colleagues who helped us prepare the camera-ready copy. For the paperback edition Padmaja Sudhakher provided significant assistance with the proofs and layout.

We have tried to identify and contact all owners of copyright material and are grateful for permission to include such material from Elsevier Science Publishers (Figures 1-12 and 7-9), Oxford University Press (Figure 4-2), Pearson Education (Figures 2-9, 3-1, 3-2, 3-3, 4-7, 6-6, 8-5, 9-1 and A2-4), Prous Science (Figure 3-4) and Springer-Verlag (Figure 1-16). We would like to offer our apologies to any copyright holders whose rights we may have unwittingly infringed.

Above all, we would like to thank those closest to us. ARL thanks Christine for once more providing the support and enthusiasm to see a book-writing project through to completion. VJG thanks Sean Paling for his boundless encouragement and support, in this, as in all things.

Chapter 1

REPRESENTATION AND MANIPULATION OF 2D MOLECULAR STRUCTURES

1. INTRODUCTION

Many organisations maintain databases of chemical compounds. Some of these databases are publicly accessible; others are proprietary. They may contain a very large number of compounds; several hundred thousand is common, and some hold millions of entries. It is nevertheless possible to search for structures and data in these databases and have the results returned within seconds. A recent development involves the creation of databases containing *virtual* molecules. These are compounds that do not exist yet (so far as the creator is aware) but which could be synthesised readily, often using combinatorial chemistry techniques. The sizes of these virtual collections can run to billions of structures.

In this chapter we introduce some of the methods used for the computational representation of molecular structures and the creation of structural databases. The procedures used to accomplish common tasks such as substructure searching are considered, including the use of graph theoretic methods, bitstring screens and the use of canonical structure representations. Finally, we give a brief overview of reaction databases and the use of Markush representations in chemical patents. The focus of this chapter will be on "2D representations" that are concerned with the chemical bonding between atoms rather than their 3D structures (the subject of Chapter 2).

2. COMPUTER REPRESENTATIONS OF CHEMICAL STRUCTURES

Let us start by considering how we might store information about a chemical structure within a computer. We will use aspirin as our example (see Figure 1-1). Easy-to-use programs are available that enable chemists to draw such structures for incorporation into reports, publications or presentations; two popular examples are ChemDraw [ChemDraw 2002] and ISIS/Draw [ISIS/Draw 2002]. One way to store the structure would be as an image, as might for example be obtained by

1

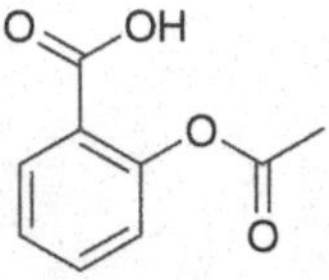

Figure 1-1. The chemical structure of aspirin (acetyl salicylic acid).

scanning a printout of the structure into the computer, or in a text format such as Portable Document Format (pdf) [pdf]. However, an image file has little real value in chemoinformatics and computational chemistry. Whilst it does enable the structure to be reproduced, a more appropriate representation is required if the molecule is to be stored in a database for subsequent retrieval and search based on its chemical structure.

2.1   Graph Theoretic Representations of Chemical Structures

Chemical structures are usually stored in a computer as *molecular graphs*. Graph theory is a well-established area of mathematics that has found application not just in chemistry but in many other areas, such as computer science. A *graph* is an abstract structure that contains *nodes* connected by *edges*. Some examples of graphs that correspond to chemical structures are shown in Figure 1-2. In a molecular graph the nodes correspond to the atoms and the edges to the bonds. Note that hydrogen atoms are often omitted. The nodes and edges may have properties associated with them. For example, the atomic number or atom type may be associated with each node and the bond order with each edge. These atom and bond properties are important when performing operations with or upon the molecular graph. A graph represents the *topology* of a molecule only, that is, the way the nodes (or atoms) are connected. Thus a given graph may be drawn in many different ways and may not obviously correspond to a "standard" chemical diagram (e.g. the graph representation of aspirin in Figure 1-2). A *subgraph* is a subset of the nodes and edges of a graph; thus the graph for benzene is a subgraph of the graph for aspirin. A *tree* is a special type of graph in which there is just a single path connecting each pair of vertices, that is, there are no cycles or rings within the graph. The *root node* of a tree is the starting point; the other vertices are either *branch nodes* or *leaf nodes*. Acyclic molecules are represented using trees. In a *completely connected graph* there is an edge between all pairs of nodes. Such graphs are rather rare in chemistry, one example being the P_4 form of elemental phosphorus, but they can be useful when manipulating 3D structures as will be seen in Chapter 2.

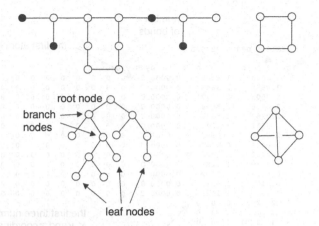

Figure 1-2. A graph consists of nodes connected by edges. The graph on the bottom left is an example of a tree and that at the bottom right is a completely connected graph. Clockwise, from top left, these graphs represent aspirin, cyclobutane, the P_4 form of phosphorus and 2,3,4,6-tetramethyloctane with C_5 as the root node. In the aspirin graph the solid nodes represent oxygen atoms.

As graphs can be constructed in many different ways it is necessary to have methods to determine whether two graphs are the same. In graph theoretic terms this problem is known as *graph isomorphism*. There are well-developed algorithms from computer science that can be used to establish whether two molecular graphs are isomorphic [Read and Corneil 1977; Trinajstic 1983]. A trivial requirement is that they must have the same number of nodes and edges; it is then necessary to establish a *mapping* from one graph to the other such that every node and edge in one graph has an equivalent counterpart in the other graph.

2.2 Connection Tables and Linear Notations

It is necessary to have a means to communicate the molecular graph to and from the computer. This can be achieved in a variety of ways. A common method is to use a *connection table*. The simplest type of connection table consists of two sections: first, a list of the atomic numbers of the atoms in the molecule and second a list of the bonds, specified as pairs of bonded atoms. More detailed forms of connection table include additional information such as the hybridisation state of each atom and the bond order. One key point to note is that hydrogen atoms may not necessarily be explicitly included in the connection table (i.e. they may be implied) in which case the connection table is *hydrogen-suppressed*. Information about the (*xy*) or (*xyz*) coordinates of the atoms may also be present to enable a standard chemical drawing to be produced or for use in a molecular graphics program. By way of example, the connection table for aspirin in the connection

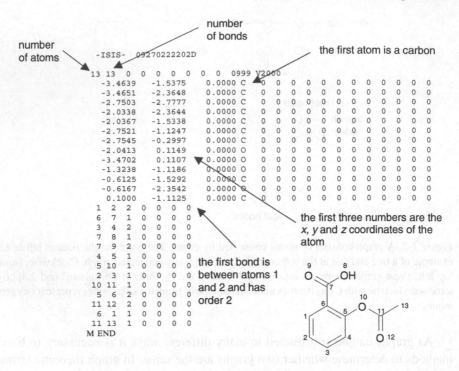

Figure 1-3. The connection table for aspirin in the MDL format (hydrogen-suppressed form). The numbering of the atoms is as shown in the chemical diagram.

table format developed by MDL Information Systems [Dalby et al. 1992] is shown in Figure 1-3. This particular format can encode a variety of information including stereochemistry, charge and isotope data.

An alternative way to represent and communicate a molecular graph is through the use of a *linear notation*. A linear notation uses alphanumeric characters to encode the molecular structure. Linear notations are more compact than connection tables and so they can be particularly useful for storing and transmitting large numbers of molecules. An early line notation that became quite widely used was the Wiswesser Line Notation (WLN) [Wiswesser 1954]. This uses a complex set of rules to represent different functional groups and the way they are connected in the molecule. A recent linear notation that has found widespread acceptance is the Simplified Molecular Input Line Entry Specification (SMILES) notation [Weininger 1988]. One reason for the extensive use of SMILES is that it is much easier to use and comprehend than the WLN; just a few rules are needed to write and understand most SMILES strings.

In SMILES, atoms are represented by their atomic symbol. Upper case symbols are used for aliphatic atoms and lower case for aromatic atoms. Hydrogen atoms are not normally explicitly represented as SMILES is a hydrogen-suppressed notation (see below for some of the exceptions). Double bonds are written using "=" and triple bonds using "#"; single and aromatic bonds are not explicitly represented by any symbol (except in special cases such as the non-aromatic single bond in biphenyl, which is written using "−"). To construct a SMILES string one needs to "walk" through the chemical structure in such a way that all the atoms are visited just once. Rings are dealt with by "breaking" one of the bonds in each ring; the presence of the ring is then indicated by appending an integer to the two atoms of the broken bond. As one progresses through the molecule one will usually arrive at branch points, where more than one possible avenue could be followed. The presence of a branch point is indicated using a left-hand bracket; a right-hand bracket indicates that all the atoms in that branch have been visited. Branches can be nested to any level necessary.

The simplest SMILES is probably that for methane: C. Note that all four attached hydrogens are implied. Ethane is CC, propane is CCC and 2-methyl propane is CC(C)C (note the branch point). Cyclohexane illustrates the use of ring closure integers; the SMILES is C1CCCCC1. Benzene is c1ccccc1 (note the use of lower case to indicate aromatic atoms). Acetic acid is CC(=O)O. The SMILES for a selection of more complex molecules are provided in Figure 1-4.

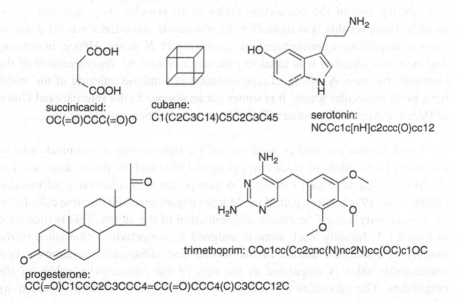

succinicacid:
OC(=O)CCC(=O)O

cubane:
C1(C2C3C14)C5C2C3C45

serotonin:
NCCc1c[nH]c2ccc(O)cc12

trimethoprim: COc1cc(Cc2cnc(N)nc2N)cc(OC)c1OC

progesterone:
CC(=O)C1CCC2C3CCC4=CC(=O)CCC4(C)C3CCC12C

Figure 1-4. Some examples of the SMILES strings for a variety of molecules.

Information about chirality and geometrical isomerism can also be included in the SMILES notation. The absolute stereochemistry at chiral atoms is indicated using the "@" symbol; by way of example, the two stereoisomers of alanine are written N[C@H](C)C(=O)O and N[C@@H](C)C(=O)O. Note that in this case the hydrogen atom on the chiral carbon is included as part of the specification of the chiral centre. Geometrical (E/Z or *cis–trans*) isomerism about double bonds is indicated using slashes; again by way of example *trans*-butene is C/C=C/C and *cis*-butene is C/C=C\C. Further information about these and other extensions to the basic SMILES notation can be found elsewhere [James et al. 2002].

2.3   Canonical Representations of Molecular Structures

There may be many different ways to construct the connection table or the SMILES string for a given molecule. In a connection table one may choose different ways to number the atoms and in SMILES notation the SMILES string may be written starting at a different atom or by following a different sequence through the molecule. For example, many possible SMILES strings can be written for aspirin, two of which are OC(=O)c1ccccc1OC(=O)C and c1cccc(OC(=O)C)c1C(=O)O). Chemical database systems need to be able to establish whether two connection tables or two SMILES strings represent the same chemical structure or not (as might arise should there be an attempt to enter the same compound twice). This problem could in principle be tackled by renumbering one of the connection tables in all possible ways and testing for identity. However this is computationally unfeasible since there are $N!$ different ways of numbering a connection table consisting of N atoms. Hence, in order to deal with this situation it is usual to generate a *canonical representation* of the chemical structure. A canonical representation is a unique ordering of the atoms for a given molecular graph. It is somewhat analogous to the International Union of Pure and Applied Chemistry (IUPAC) name for a chemical structure.

A well-known and widely used method for determining a canonical order of the atoms is the Morgan algorithm [Morgan 1965] and its descendants such as the SEMA method, which extends it to incorporate stereochemistry information [Wipke et al. 1974]. A key part of the Morgan algorithm is the iterative calculation of "connectivity values" to enable differentiation of the atoms. This is illustrated in Figure 1-5. Initially, each atom is assigned a connectivity value equal to the number of connected atoms. In the second and subsequent iterations a new connectivity value is calculated as the sum of the connectivity values of the neighbours. The procedure continues until the number of different connectivity

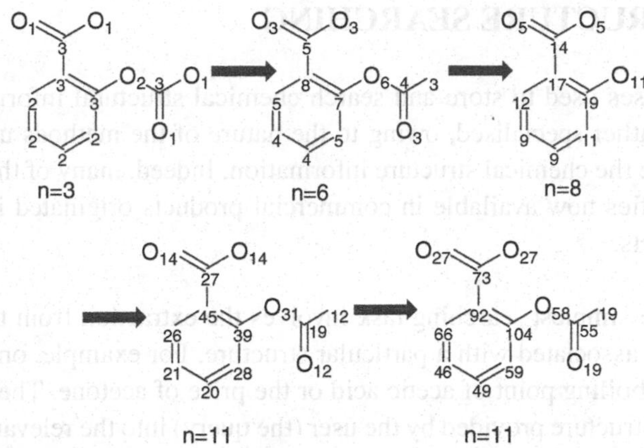

Figure 1-5. Illustration of the way in which the Morgan algorithm iteratively constructs the atomic connectivity values.

values reaches a maximum. Thus, in the case of aspirin there are initially three different connectivity values (1, 2, 3). This increases in the subsequent iterations to six, eight and finally eleven. The atom with the highest connectivity value is then chosen as the first atom in the connection table; its neighbours are then listed (in order of their connectivity values), then their neighbours and so on. If a "tie" occurs then additional properties are considered such as atomic number and bond order. For example, the two oxygens of the carboxylic acid in aspirin have the same connectivity value, as do the terminal methyl and the carbonyl oxygen of the acetyl group. These conflicts can be resolved by consideration of the bond order and atomic number respectively.

An algorithm called CANGEN has been developed to generate a unique SMILES string for each molecule [Weininger et al. 1989] (the canonical SMILES for aspirin is CC(=O)Oc1ccccc1C(=O)O). This uses somewhat similar principles to the Morgan algorithm to produce a canonical ordering of the atoms in the graph. This canonical ordering is then used to generate the unique SMILES string for the molecule.

Many other formats have been devised for representing molecular graphs. Two recent examples that are designed to have general utility are the canonical InChI representation (formerly IChI) being developed by IUPAC [InChI; Stein et al. 2003] and the Chemical Markup Language (CML) that is designed for the exchange of information over the Internet [Murray-Rust and Rzepa 1999].

3.   STRUCTURE SEARCHING

The databases used to store and search chemical structural information have tended to be rather specialised, owing to the nature of the methods used to store and manipulate the chemical structure information. Indeed, many of the chemistry database facilities now available in commercial products originated in academic research projects.

Perhaps the simplest searching task involves the extraction from the database of information associated with a particular structure. For example, one may wish to look up the boiling point of acetic acid or the price of acetone. The first step is to convert the structure provided by the user (the query) into the relevant canonical ordering or representation. One could then search through the database, starting at the beginning, in order to find this structure. However, the canonical representation can also provide the means to retrieve information about a given structure more directly from the database through the generation of a *hash key*. A hash key is typically an integer with a value between 0 and some large number (e.g. $2^{32} - 1$). If the database is arranged so that the hash key corresponds to the physical location on the computer disk where the data associated with that structure is stored, then the information can be retrieved almost instantaneously by moving the disk read mechanism directly to that location. There are well-established computer algorithms for the generation of hash keys [Wipke et al. 1978]. Much attention in particular has been paid to the generation of hash keys for strings of characters (such as canonical SMILES strings). An example of a hash key devised specifically for chemical information is the Augmented Connectivity Molecular Formula used in the Chemical Abstracts Service (CAS) Registry System [Freeland et al. 1979] which uses a procedure similar in concept to the Morgan algorithm. Ideally, each canonical structure will produce a different hash key. However, there is always a chance that two different structures will produce the same hash key. It is thus necessary to incorporate mechanisms that can automatically resolve such clashes.

4.   SUBSTRUCTURE SEARCHING

Beyond looking up the data associated with a particular structure, *substructure searching* is perhaps the most widely used approach to identify compounds of interest. A substructure search identifies all the molecules in the database that contain a specified substructure. A simple example would be to identify all structures that contain a particular functional group or sequence of atoms such as a carboxylic acid, benzene ring or C_5 alkyl chain. An illustration of the range of hits that may be obtained following a substructure search based on a dopamine derivative is shown in Figure 1-6.

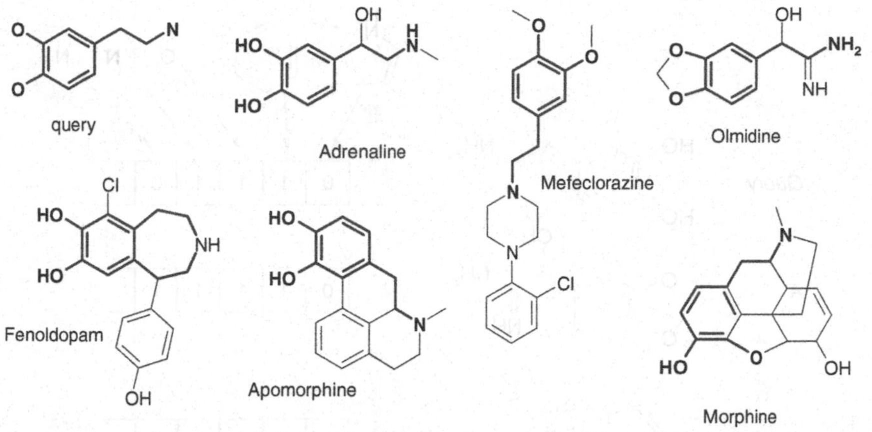

Figure 1-6. An illustration of the range of hits that can be obtained using a substructure search. In this case the dopamine-derived query at the top left was used to search the World Drug Index (WDI) [WDI]. Note that in the case of fenoldopam the query can match the molecule in more than one way.

Graph theoretic methods can be used to perform substructure searching, which is equivalent to determining whether one graph is entirely contained within another, a problem known as *subgraph* isomorphism. Efficient algorithms for performing subgraph isomorphism are well established, but for large chemical databases they are usually too slow to be used alone. For this reason most chemical database systems use a two-stage mechanism to perform substructure search [Barnard 1993]. The first step involves the use of *screens* to rapidly eliminate molecules that cannot possibly match the substructure query. The aim is to discard a large proportion (ideally more than 99%) of the database. The structures that remain are then subjected to the more time-consuming subgraph isomorphism procedure to determine which of them truly do match the substructure. Molecule screens are often implemented using binary string representations of the molecules and the query substructure called *bitstrings*. Bitstrings consist of a sequence of "0"s and "1s". They are the "natural currency" of computers and so can be compared and manipulated very rapidly, especially if held in the computer's memory. A "1" in a bitstring usually indicates the presence of a particular structural feature and a "0" its absence. Thus if a feature is present in the substructure (i.e. there is a "1" in its bitstring) but not in the molecule (i.e. the corresponding value is "0") then it can be readily determined from the bitstring comparison that the molecule cannot contain the substructure. The converse does not hold; there will usually be features present in the molecule that are not in the substructure. An example of the use of bitstrings for substructure search is shown in Figure 1-7.

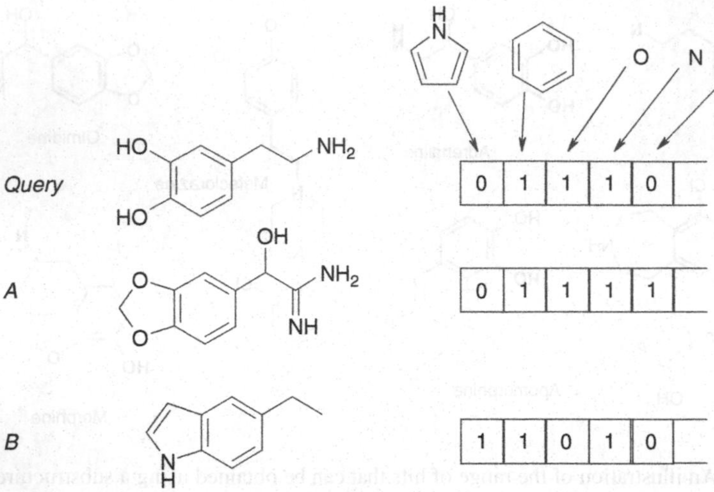

Figure 1-7. The bitstring representation of a query substructure is illustrated together with the corresponding bitstrings of two database molecules. Molecule *A* passes the screening stage since all the bits set to "1" in the query are also set in its bitstring. Molecule *B*, however, does not pass the screening stage since the bit representing the presence of O is set to "0" in its bitstring.

4.1   Screening Methods

Most binary screening methods use one of two approaches. In a *structural key*, each position in the bitstring corresponds to the presence or absence of a predefined substructure or molecular feature. This is often specified using a *fragment dictionary*; the *i*th substructure in the dictionary corresponds to the *i*th bit in the bitstring. When a new molecule is added to the database each of the fragments in the dictionary is checked against it. If the fragment is present then the relevant bit is set to "1" (the bitstring initially has all positions set to "0"). In addition to the presence or absence of substructures such as functional groups and rings, a structural key may contain information about particular elements and about the number of a particular type of feature (e.g. a particular bit may correspond to "at least two methyl groups"). The aim when constructing a fragment dictionary is to define a structural key that gives optimal performance in a typical search, by eliminating a large proportion of the database prior to the subgraph isomorphism stage. It is thus necessary to design the dictionary according to the molecules expected to be stored in the database and also to consider typical queries that might be submitted. For example, the keys used for a database of typical organic molecules will probably be inappropriate for a database of solid-state materials or a database consisting entirely of hydrocarbons. There has been much work to determine the most effective set of substructures to use in the keys [Adamson et al. 1973; Hodes 1976], and automated methods for screen selection have been developed that are based on the statistical occurrence of substructures in the database [e.g. Willett 1979]. Typically the aim is to select fragments that are independent of each other and that are equifrequent since fragments that are very frequent and occur in most molecules are unlikely to be discriminating and conversely fragments that occur infrequently are unlikely to be useful. Thus, frequently occurring fragments are discarded and infrequently occurring fragments are generalised or grouped to increase their frequency.

In principle any fragment is possible, but certain types are frequently encountered. For example, versions of the CAS Registry System use twelve different types of screens [Dittmar et al. 1983]. These include: augmented atoms (consisting of a central atom together with its neighbouring atoms and bonds); atom sequences (linear sequences of connected atoms with or without the bond types differentiated); bond sequences (atom sequences with the bonds differentiated but not the atoms); and screens associated with ring systems such as the number of rings in a molecule and the ring type (which indicates the sequence of connectivities around a ring). These are illustrated in Figure 1-8. Of those systems which use structural key and a fragment dictionary the MACCS and ISIS systems from MDL Information Systems Inc. [MDL] and the BCI fingerprints from Barnard Chemical Information Ltd. [BCI] are widely used in chemoinformatics.

The alternative to structural keys is to use a *hashed fingerprint*, which does not require a predefined fragment dictionary, and so in principle has the advantage of being applicable to any type of molecular structure. It is produced by generating all

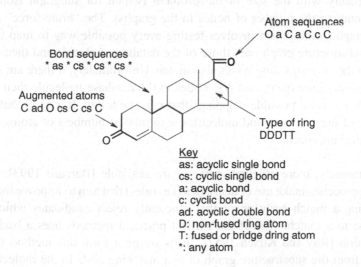

Atom sequences
O a C a C c C

Bond sequences
* as * cs * cs *

Augmented atoms
C ad O cs C cs C

Type of ring
DDDTT

Key
as: acyclic single bond
cs: cyclic single bond
a: acyclic bond
c: cyclic bond
ad: acyclic double bond
D: non-fused ring atom
T: fused or bridge dring atom
*: any atom

Figure 1-8. Examples of the types of substructural fragments included in screening systems (in bold).

possible linear paths of connected atoms through the molecule containing between one and a defined number of atoms (typically seven). For example, when using the SMILES representation for aspirin the paths of length zero are just the atoms c, C and O; the paths of length one are cc, cC, C=O, CO and cO, and so on. Each of these paths in turn serves as the input to a second program that uses a hashing procedure to set a small number of bits (typically four or five) to "1" in the fingerprint bitstring. A feature of the hashing method is that each bit in the fingerprint can be set by more than one pattern (resulting in a *collision*) though it is much less likely (though in principle still possible) that all of the bits set by one pattern will also be set by a different pattern. The presence of collisions in hashed fingerprints does not result in false negatives (wherein structures that match the query would be missed) but they can result in more molecules having to be considered for the slower subgraph isomorphism stage. Hashed fingerprints are particularly associated with Daylight Chemical Information Systems Inc. [Daylight] where typical organic molecules set between 50 and 400 bits (out of a total length of 1,024). Some systems use a combination of structural keys and hashed fingerprints (e.g. the UNITY system [UNITY]).

4.2   Algorithms for Subgraph Isomorphism

Having eliminated molecules using the bitstring screen the subgraph isomorphism search is then performed on the molecules that remain to determine which actually match the substructure. The subgraph isomorphism search is inherently slower; indeed it belongs to a class of problems known as *NP-complete*, meaning that to find a solution all algorithms require an amount of time that varies exponentially with the size of the problem (which for subgraph isomorphism corresponds to the number of nodes in the graphs). The "brute-force" approach to subgraph isomorphism involves testing every possible way to map the nodes of the substructure graph onto those of the database molecule and then checking whether the corresponding bonds also match. Unfortunately, if there are N_S nodes in the substructure query and N_M nodes in the database molecule then there are $N_M!/(N_M - N_S)!$ possible mappings that must be tested. It is clear that even for very small substructures and molecules with modest numbers of atoms this is an impractical approach.

Fortunately, more efficient methods are available [Barnard 1993]. Many of these approaches make use of heuristics (i.e. rules) that aim to improve the chances of finding a match early on or that efficiently reject candidates which cannot give rise to a match. One of the earliest practical methods uses a backtracking mechanism [Ray and Kirsch 1957]. In its simplest form this method first maps a node from the substructure graph onto a matching node in the molecule graph (such that, for example, the two matching nodes have the same atom type). The algorithm next attempts to map neighbouring nodes of the substructure node onto the corresponding neighbouring nodes in the molecule graph. This process continues until all the nodes have been successfully matched or until it proves impossible to find a match, at which point the algorithm returns (or *backtracks*) to the last successful match and attempts an alternative mapping. This backtracking procedure continues until either a match of all nodes can be found or until all possible mappings of the initial node have been attempted, at which point the algorithm terminates because the substructure cannot be present in the molecule. To improve the efficiency of this algorithm further it is possible to take into account additional characteristics of the graph nodes such as information about a node's neighbours. This may enable unproductive mappings to be rejected even earlier in the process.

The Ullmann algorithm [Ullmann 1976] is a widely used method for subgraph isomorphism that has been shown to be one of the most efficient algorithms for chemical structures [Downs et al. 1988a; Brint and Willett 1987]. It uses *adjacency matrices* to represent the molecular graphs of the query substructure and the molecule. In an adjacency matrix the rows and columns correspond to the atoms in the structure such that if atoms i and j are bonded then the elements (i,j) and (j,i) of the matrix have the value "1". If they are not bonded then the value is assigned "0". It is thus an alternative formulation of the information in a connection table. The adjacency matrix for aspirin is shown in Figure 1-9.

The Ullmann algorithm operates as follows. Suppose there are N_S atoms in the substructure query and N_M atoms in the database molecule and that **S** is the adjacency matrix of the substructure and **M** is the adjacency matrix of the molecule. A matching matrix **A** is constructed with dimensions N_S rows by N_M

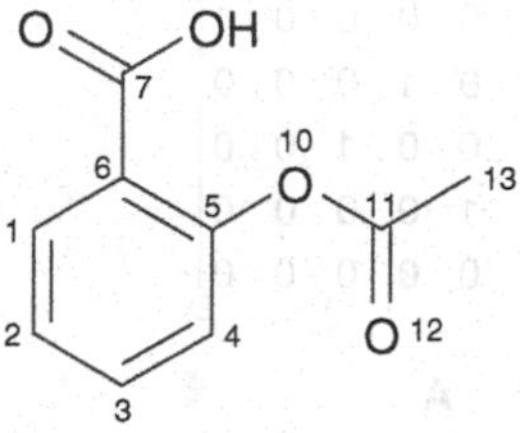

$$
\begin{pmatrix}
0 & 1 & 0 & 0 & 0 & 1 & 0 & 0 & 0 & 0 & 0 & 0 & 0 \\
1 & 0 & 1 & 0 & 0 & 0 & 0 & 0 & 0 & 0 & 0 & 0 & 0 \\
0 & 1 & 0 & 1 & 0 & 0 & 0 & 0 & 0 & 0 & 0 & 0 & 0 \\
0 & 0 & 1 & 0 & 1 & 0 & 0 & 0 & 0 & 0 & 0 & 0 & 0 \\
0 & 0 & 0 & 1 & 0 & 1 & 0 & 0 & 0 & 1 & 0 & 0 & 0 \\
1 & 0 & 0 & 0 & 1 & 0 & 1 & 0 & 0 & 0 & 0 & 0 & 0 \\
0 & 0 & 0 & 0 & 0 & 1 & 0 & 1 & 1 & 0 & 0 & 0 & 0 \\
0 & 0 & 0 & 0 & 0 & 0 & 1 & 0 & 0 & 0 & 0 & 0 & 0 \\
0 & 0 & 0 & 0 & 0 & 0 & 1 & 0 & 0 & 0 & 0 & 0 & 0 \\
0 & 0 & 0 & 0 & 1 & 0 & 0 & 0 & 0 & 0 & 1 & 0 & 0 \\
0 & 0 & 0 & 0 & 0 & 0 & 0 & 0 & 0 & 1 & 0 & 1 & 1 \\
0 & 0 & 0 & 0 & 0 & 0 & 0 & 0 & 0 & 0 & 1 & 0 & 0 \\
0 & 0 & 0 & 0 & 0 & 0 & 0 & 0 & 0 & 0 & 1 & 0 & 0
\end{pmatrix}
$$

Figure 1-9. The adjacency matrix for aspirin, using the atom numbering shown.

columns so that the rows represent the atoms of the substructure and the columns represent the atoms of the database molecule. The elements of the matching matrix take the value "1" if a match is possible between the corresponding pair of atoms and "0" otherwise. The aim is to find matching matrices in which each row contains just one element equal to "1" and each column contains no more than one element equal to "1". In this case a unique mapping has been found for every atom of the substructure and for those elements A_{ij} that have a value "1", atom i in the substructure matches atom j in the molecule. More concisely, each matching matrix must satisfy $\mathbf{A}(\mathbf{AM})^T = \mathbf{S}$ (the superscript^T indicates the transposed matrix in which the rows and columns of the matrix have been interchanged; a summary of matrix operations and properties can be found in Appendix 1). By way of illustration, Figure 1-10 shows such a situation with one of the possible matches of the substructure onto the molecule and the corresponding matrices.

One way to implement the Ullmann algorithm would be to generate systematically all possible matrices $\mathbf{A}$, testing them to determine whether they

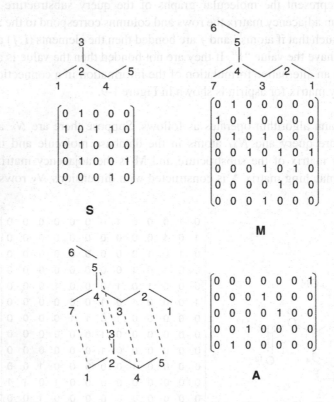

Figure 1-10. Simple illustration of the operation of the Ullmann algorithm using a 5-atom substructure and a 7-atom "molecule" represented by their adjacency matrices. One of the proposed matches is shown in the bottom figure together with the corresponding matching matrix.

correspond to a viable match. However, this would be a rather time-consuming and inefficient procedure and so more effective methods have been devised. Ullmann himself suggested that neighbouring atoms could be taken into account when considering possible matches (i.e. an atom in the query is not permitted to match an atom in the molecule unless each of their neighbouring atoms also match). This step in which an initial match is improved by using local information is referred to as a *refinement* or *relaxation* step. Atom matches are initially assigned in the matching matrix based on atom type and connectivity. A depth-first search is then carried out incorporating the refinement step suggested above. Thus, the first substructure atom is initially assigned to the first molecule atom that it might match, the appropriate column in the matching matrix is set to "1" and all remaining columns in that row are set to "0". The refinement step is then applied; if it fails then the algorithm backtracks and the next potential match is tested. If it succeeds then the next row of the substructure is operated on. The process is repeated recursively until either a complete match for the substructure is discovered or until all options have been attempted. The refinement step and related techniques enable current implementations of the Ullmann algorithm to provide very efficient performance even with today's large databases.

Different approaches to the substructure search problem have been adopted by some systems that involve preprocessing the entire database into a format that then enables the searches to be performed extremely rapidly. For example, in the approach described by Christie et al. [1993] the screening stage of substructure search is implemented using *inverted files*. A set of bitstrings is constructed, one for each fragment included in a structural key. These bitstrings are of length equal to the number of molecules in the database, with the ith bit being set if the ith molecule contains the fragment. For molecule i to be a potential match the ith bit must be set to "1" in the bitstrings of all the fragments in the substructure. This can be determined by combining the relevant bitstrings using the logical **AND** operator. The ith position in **A AND B** is set to "1" only if this bit is set in both of the bitstrings **A** and **B**; otherwise it is set to "0". These inverted bitstrings provided a more rapid way to identify potential matches than traditional bitstrings. More complex systems have also been invented that involve encoding the atoms in each molecule in a way that reflects their environments (e.g. the type of the atom and its bonds, the number of neighbours, whether it is in a ring or in an acyclic chain). As this encoding is extended to the second and third neighbours of each atom, so the fragments become more and more discriminating. In the Hierarchical Tree Substructure Search (HTSS) system [Nagy et al. 1988] the entire database is stored in a tree-like structure based on these encoded atoms. A substructure search on the entire database is then reduced to a set of simple tree searches. Other database systems use a similar approach, including the S4 system [Hicks and Jochum 1990] developed by the Beilstein Institute and its collaborators. One potential drawback of some of these systems is that it can be difficult to update the database with new structures. This may not be a problem for databases that are released on a periodic basis, but can be difficult for systems that need to be continually updated.

4.3   Practical Aspects of Structure Searching

Modern substructure search methods enable very complex queries to be constructed. For example, one may wish to specify that an atom is one of a group (e.g. "can match any halogen") and/or that an atom should *not* be a specific atom (e.g. "oxygen or carbon but not nitrogen"). It may be desired to specify that certain atoms or bonds must be part of a ring (or must not), that atoms must have hydrogens attached (or not), and so on. The need to formulate such queries arises from the complex nature of chemistry and the ways in which molecules interact with each other and their environment. The MDL connection table provides special fields to specify substructure-related information [Dalby et al. 1992] and SMARTS [James et al. 2002] is an extension of the SMILES language for substructure specification.

The analysis and manipulation of chemical structures often involves the determination of features such as bond orders, rings and aromaticity. This information may be used to generate structural keys or hashed fingerprints, and may also be used to make the subgraph isomorphism step more efficient (e.g. if the substructure contains a ring then these atoms must match an equivalent ring in the molecule). In some cases these features are explicit in the representation provided by the user; in other cases they have to be inferred from the information provided. Ring systems in particular play a key role in chemistry and so efficient methods for the perception of rings present in a molecular graph have attracted much attention over many years. When bridged and fused ring systems are present then there may be a number of different ways in which the rings can be classified. Many methods have been proposed for finding the ring systems in molecular graphs [Downs et al. 1989]. One approach is to identify all of the rings. Thus for example in decalin one would identify the two 6-membered rings together with the encompassing 10-membered ring. A common alternative is to identify the so-called Smallest Set of Smallest Rings (SSSR). This is a set of rings from which all others in the molecular graph can be constructed. Moreover, it comprises those rings containing the fewest atoms. Thus, for example, in decalin the SSSR contains two 6-membered rings. The SSSRs for some ring systems are shown in Figure 1-11. Having identified the rings present it is then possible to identify those that are potentially aromatic; this is usually done on the basis of the Hückel $4n + 2$ rule.

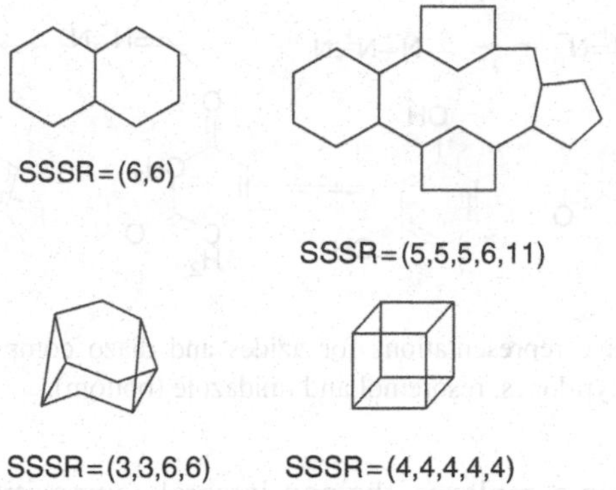

SSSR=(6,6)

SSSR=(5,5,5,6,11)

SSSR=(3,3,6,6)   SSSR=(4,4,4,4,4)

Figure 1-11. Some ring systems and the SSSRs that they contain.

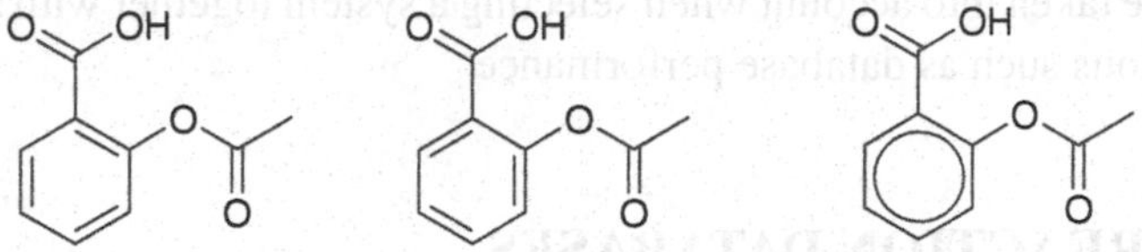

Figure 1-12. Alternative ways to represent the aromatic ring in aspirin.

Several practical issues must be considered when working with chemical structure representations and databases [Barnard 1988]. One of these concerns the representation of aromatic systems. For example, three different, but equivalent representations of aspirin are shown in Figure 1-12. Each of these would be considered perfectly valid by a chemist. Moreover, he or she would expect any computer system to be able to automatically recognise the equivalence of these three representations.

In addition to aromatic systems certain other functional groups may be represented in more than one way. Common examples include azides and diazo compounds (Figure 1-13). Another feature to be considered is tautomerism. In many cases the dominant tautomer can be easily identified and would be the representation present in the database (e.g. acetone only contains approximately 0.0001% of the enol form). However, in other systems there may be more than one viable tautomer, classic examples being pyridones, resorcinol and imidazoles (Figure 1-13).

In addition to the organic compounds that have been the main focus of our discussion it may also be important to be able to represent inorganic molecules, coordination complexes, macromolecules (such as proteins or DNA) and polymers. Each of these presents challenges, such as the complex stereochemistry

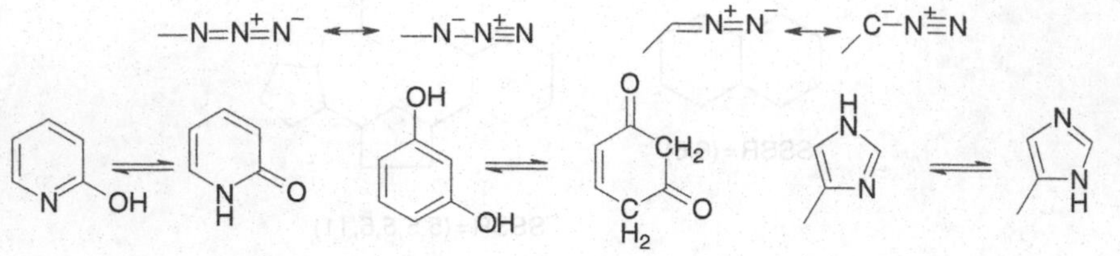

Figure 1-13. Alternative representations for azides and diazo compounds (top). Alternative tautomeric forms for pyridones, resorcinol and imidazole (bottom).

of some coordination complexes, the non-integral composition of some inorganic compounds and the sheer number of atoms in macromolecules or polymers. Few systems are designed to cope with all possibilities; these factors should therefore be taken into account when selecting a system together with any practical considerations such as database performance.

5.   REACTION DATABASES

Reactions are central to the subject of chemistry, being the means by which new chemical entities are produced. As any student of chemistry is aware a huge number of new reactions and new applications of existing reactions are published every year. It is clearly important to be able to take published reaction information into account when planning a new chemical synthesis. This is particularly important as a chemist may spend months refining and optimising a particular synthetic procedure; once this is done then other scientists need to be able to benefit from this knowledge. Traditionally, of course, this was achieved via the printed chemical literature. Summaries of the primary literature such as *The Beilstein Handbuch der Organischen Chemie* (first published in 1881 and containing information from 1771 onwards [Luckenbach 1981]) have proved to be invaluable aids for chemists wishing to plan their own syntheses. The potential of computational methods for storing, searching and retrieving information about chemical reactions has long been recognised [Lynch and Willett 1978]. This has resulted in both commercial and in-house reaction database systems that enable valuable information to be captured for subsequent dissemination and reuse.

When planning a synthesis a chemist may wish to search a reaction database in a variety of ways. A simple type of query would involve an exact structure search against the products, to determine whether there was an established synthesis for the compound of interest. In the case of a novel compound there may be related structures with precedented synthetic routes. In other cases the chemist may be more concerned with functional group transformations. For example, one may wish to identify potential routes to aminothiazoles. Sometimes it will be desired to perform a *reaction search* involving the structures or substructures of the precursor or reactant and the product. A more general type of search would involve identification of all deposited syntheses for a particular named reaction, in order to identify a range of possible reaction conditions to attempt. Other quantities of interest may include specific reagents, catalysts, solvents, the reaction conditions (temperature, pressure, pH, time, etc.) together with the yield. One may of course wish to combine more than one of these criteria in a single query (e.g. "find all deposited reaction schemes which involve the reduction of an aliphatic aldehyde to an alcohol where the temperature is less than 150°C and the yield is greater than 75%").

As might be expected from our discussion above, the design of reaction database systems is a very complex task if all of the above criteria are to be satisfied. The earliest reaction database systems were purely text-based, enabling queries for information such as authors, compound name and compound formula via the use of keywords. Subsequent developments provided the ability to incorporate structural information into the search query. Some current systems provide the ability to perform exact and substructure searches against reaction products, thereby retrieving all reactions which contain the key fragment in a product molecule. Other systems provide reaction searching, in which one can define structural features of both reactant(s) and product(s), in order to retrieve reactions that involve the specific transformation. A key component of such reaction searching is the use of a procedure known as *atom mapping* [Lynch and Willett 1978; Chen et al. 2002]. This enables the 1:1 correspondences between atoms in the reactants and products to be defined. An example is shown in Figure 1-14. The atom mapping of a reaction may also provide information about the underlying mechanism, including changes in stereochemistry. When used as part of a reaction query it enables the system to retrieve just those reactions in which the specified substructure in the reactant(s) is converted to the specified substructure in the product(s) and no others.

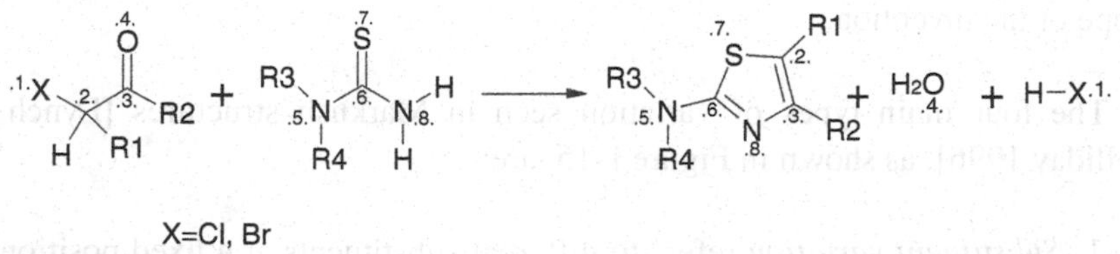

Figure 1-14. An example of atom mapping, in this case for the formation of aminothiazoles from thioureas and α-haloketones.

Given the volume of published material, it is a significant task to maintain those databases which survey all or a significant proportion of the chemical literature. For example, the Beilstein database contains more than 9 million reactions [Meehan and Schofield 2001; Crossfire] and the CASREACT database more than 6 million reactions [Blake and Dana 1990; CASREACT]. Other databases do not aim to include all of the primary literature, but rather to include a representative set of examples of each reaction to provide a chemist with sufficient material to tackle their particular problem. There have been some published comparisons of the different reaction database systems [Borkent et al. 1988; Zass 1990; Parkar and Parkin 1999; Hendrickson and Zhang 2000]. In addition to the major efforts expended implementing and populating reaction databases a number of research groups continue to investigate novel computational algorithms both for the representation of chemical reactions and as aids to assist in the design of chemical syntheses.

6. THE REPRESENTATION OF PATENTS AND PATENT DATABASES

The patent system confers a period of exclusivity and protection for a novel, useful and non-obvious invention in exchange for its disclosure in the form of a patent specification. Patents are of great importance in the chemical and pharmaceutical industries with many companies competing in the same field, frequently on the same or a similar series of molecules. It is therefore of crucial importance to be able to determine whether a particular molecule or series of molecules is covered by one's own or someone else's patent. Unfortunately this cannot be achieved by simply looking up the structures in a database. This is because the scopes of chemical disclosures in patents are most frequently expressed using generic, or *Markush*, structures in which variables are used to encode more than one structure into a single representation. These representations can refer to an extremely large number of molecules. Many of the individual molecules may not actually have been synthesised or tested but are included to ensure that the coverage of the patent is sufficiently broad to embrace the full scope of the invention.

The four main types of variation seen in Markush structures [Lynch and Holliday 1996], as shown in Figure 1-15, are:

1. *Substituent variation* refers to different substituents at a fixed position, for example "R1 is methyl or ethyl".
2. *Position variation* refers to different attachment positions as shown by R2 in Figure 1-15.

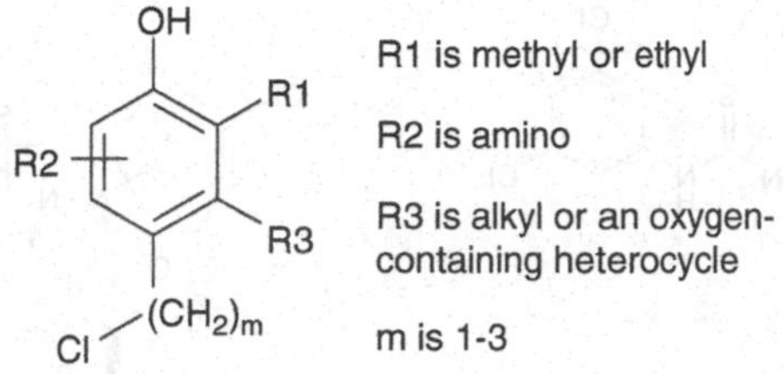

OH

R1 is methyl or ethyl

R2 is amino

R3 is alkyl or an oxygen-containing heterocycle

m is 1-3

Figure 1-15. The types of variation seen in Markush structures (see text).

3. *Frequency variation* refers to different occurrences of a substructure, for example "$(CH_2)_m$ where *m* is from 1 to 3".

4. *Homology variation* refers to terms that represent chemical families, for example "R3 is alkyl or an oxygen-containing heterocycle".

In some cases, it may be difficult, or even impossible, to explicitly enumerate all possible chemical structures that fall within the generic scope of a patent disclosure. Despite these challenges, as we have indicated, it is clearly very important to be able to determine whether a given structure or series of structures is covered by a patent. Hence special methods have been developed for the storage, search and manipulation of generic structures.

The commercially available systems MARPAT [Ebe et al. 1991] and Markush DARC [Benichou et al. 1997] are based on techniques originally developed at Sheffield University [Lynch and Holliday 1996]. These techniques include the definition of a formal and unambiguous language called GENSAL that is used to describe structures for computer input; the Extended Connection Table Representation (ECTR) for the internal representation of generic structures; and a series of search algorithms for accessing structures stored within a database. The ECTR consists of a series of partial structures that represent the individual substituents together with the logical framework that indicates how the partial structures are connected together. The search system consists of three stages. First, a fragment-screening step which is similar to that used for the substructure search of specific structures [Holliday et al. 1992]. However, in this case two sets of fragments are generated: those that are common to all specific structures represented by the generic structure and those that occur at least once. The second stage involves an intermediate level of search based on a *reduced graph* representation [Gillet et al. 1991]. Different types of reduced graph can be employed. In one form, a structure is fragmented into cyclic and acyclic components which themselves become nodes in a graph within which the relationships between the fragments are retained. This enables real fragments such as "phenyl" to be matched against homologous series identifiers such as "aryl ring". A second form of reduced graph involves the differentiation between

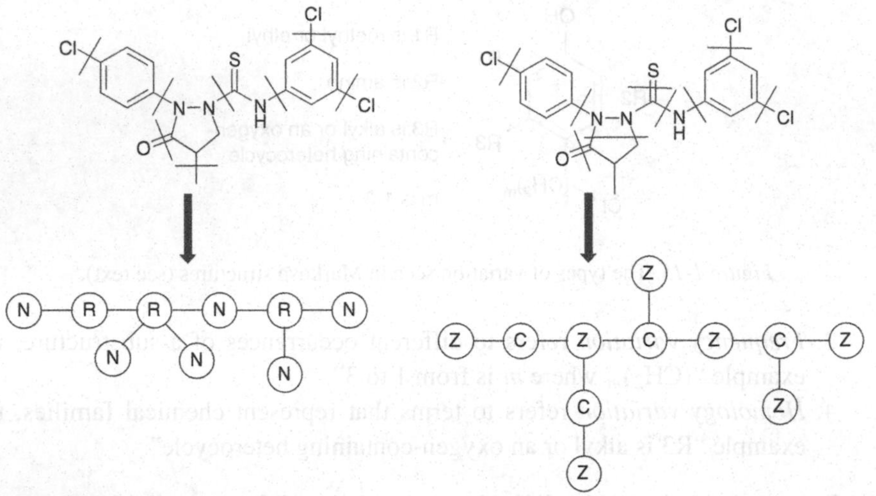

Figure 1-16. Generation of ring/non-ring (left) and carbon/heteroatom (right) forms of reduced graph. (Adapted from Downs, Gillet, Holliday and Lynch 1988.)

contiguous assemblies of carbon atoms and heteroatoms. These two forms are illustrated in Figure 1-16 [Downs et al. 1988b]. The result of a successful search at the reduced graph level is a list of pairs of matching nodes. The third and final stage of the search involves a modified version of the Ullmann algorithm which is applied on a node-by-node basis [Holliday and Lynch 1995].

7. RELATIONAL DATABASE SYSTEMS

It is rare for a database to contain just molecular structures; in almost all cases other types of data will also be present. Most databases include an identifier for each structure, such as a Chemical Abstracts Registry Number (a CAS number), an internal registry number, catalogue number or chemical name. Other types of data that may be present include measured and calculated physical properties, details of supplier and price where appropriate, date of synthesis and so on. The structure and organisation of any database is critical to the type of data that can be stored and how it is retrieved.

Relational database systems have been used for many years to store numeric and textual data [Begg and Connely 1998; Date 2000]. The principles underlying relational databases are very well understood and there are many software systems available, both commercial products as well as shareware and freeware. Until recently, however, these systems had no structure-handling capabilities and hence they have not generally been suitable for the storage of chemical information. It was therefore necessary to use separate chemical structure and relational database systems, which had to be linked in some way. The recent introduction of structure-handling capabilities (variously called *data cartridges* or *datablades*) into some of the most widely used relational database systems is likely to have a major impact on the storage, search and retrieval of chemical structure data in the near future.

In a relational database the data is stored in rectangular tables, the columns corresponding to the data items and each row representing a different piece of data. A typical database contains a number of tables, linked via unique identifiers. By way of example, suppose we wish to construct a simple relational database to hold inventory and assay data relating to biological testing. Three tables that might be used to construct such a database are shown in Figure 1-17. As can be seen, inventory information about each compound is held in one table (with relevant data, such as date synthesised and amount). Information about each assay is held in

inventory

compound_id	chemist_id	date	amount_gm	mwt
MOL1	ARL	23/08/2002	0.01	320
MOL2	ARL	23/08/2002	0.14	440
MOL3	VJG	24/08/2002	0.05	260
MOL4	VJG	25/08/2002	1.52	376
MOL5	ARL	30/08/2002	0.26	486

assay_info

assay_id	target	substrate
1	trypsin	Z-Arg
2	thrombin	D-Phe-Pro-Arg
3	chymotrypsin	Suc-Ala-Ala-Pro-Phe
4	thrombin	Tos-Gly-Pro-Arg

assay_results

compound_id	assay_id	inhibition
MOL1	1	0.5
MOL2	2	1.3
MOL1	3	0.2
MOL2	4	1.1
MOL2	3	10
MOL4	2	0.3

Figure 1-17. Three tables that might be used in a relational database to store compound registration information, assay details and assay results.

```
1. Find all compounds synthesised by chemist ARL together with their amounts available
SQL: select compound_id, amount_gm from inventory where chemist_id='ARL'
        →       MOL1   0.01
                MOL2   0.14
                MOL5   0.26

2. Find all inhibition data for compounds tested against thrombin
SQL: select compound_id, inhibition from assay_results r, assay_info a where r.assay_id=a.assay_id and
a.target='thrombin'
        →       MOL2   1.3
                MOL2   1.1
                MOL4   0.3

3. Find all compounds with thrombin inhibition less than 1.25 for which there is more than 1.0gm available
SQL: select i.compound_id, amount_gm from assay_results r, inventory i, assay_info a where r.inhibition<1.25
and r.compound_id=i.compound_id and i.amount_gm>1.0 and r.assay_id=a.assay_id and a.target='thrombin'
        →       MOL4   1.52
```

Figure 1-18. SQL queries that could be used with the tables shown in Figure 1-17.

a second table; the actual assay results themselves are stored in a third table. As can be seen, this third table is arranged so that each assay result is a separate row. An alternative would be to use one row per compound, with an entry for every assay. A drawback of this alternative database design is that it is rather inefficient because not every compound is tested in every assay. Moreover, the table would also need to be extended by adding a new column whenever a new assay is introduced. This is an example of *normalisation*. The results table in our database is linked to the inventory and assay information tables through the use of identifiers (strings of characters or numbers) that uniquely identify a particular row in the relevant table. Note for example the presence of two thrombin assays in the *assay_info* table; these are distinguished through the use of different identifiers (the *assay_id* column). It is not necessary for an identifier to have any particular meaning; for this reason they are often chosen as a sequence of integers. However, it is also possible to use textual strings as illustrated by the use of the compound identifier in this example.

A relational database is typically queried using Structured Query Language (SQL). Some SQL queries that might be used with these three tables are shown in Figure 1-18. As can be seen, in some cases it is necessary to query across multiple tables, an operation known as a *join*.

8.   SUMMARY

Methods for the representation of 2D molecular structures form the foundation for all of the techniques that we will discuss in this book. The central role played by 2D database systems is reflected in the significant amount of effort that has been expended to implement and optimise methods for the storage, search and retrieval of chemical structures and molecular data. Additional challenges are presented by reaction and patent data. Historically, chemical structural data have been stored in specialist database systems, which can lead to an unfortunate separation from other types of data. However, recent developments in database technology are helping to overcome this problem. In addition, the bitstring methods originally designed to enable efficient substructure searching have found widespread use in many other areas, such as similarity searching and the selection of diverse sets of compounds, as will be seen in later chapters.

Chapter 2

REPRESENTATION AND MANIPULATION OF 3D MOLECULAR STRUCTURES

1. INTRODUCTION

The methods described in Chapter 1 are solely concerned with 2D representations of molecules based on the molecular graph. Such representations may be considered to constitute the "natural language" of organic chemistry. However, they only indicate which atoms are bonded together. The steric and electronic properties of a molecule depend on how its atoms can be positioned in space to produce its 3D structures or *conformations*. There has thus been much interest in the development of algorithms and database systems that deal with 3D representations of molecules and their conformationally dependent properties.

There are a number of challenges associated with the use of 3D representations. Most molecules of interest can adopt more than one low-energy conformation and in many cases the number of accessible structures is very large. It is therefore necessary to have efficient ways to take conformational flexibility into account. A true representation of a molecule's properties and characteristics would require very complex computational models such as quantum mechanics or molecular simulation [Leach 2001]. These are inappropriate for dealing with large numbers of molecules and so it is necessary to develop techniques that can represent the key characteristics of a molecule's conformational properties in a computationally efficient manner.

In this chapter we will first introduce some of the concepts underlying 3D methods by describing two widely used databases that contain experimentally determined 3D data. We then consider the concept of a 3D pharmacophore. This leads to the use of 3D databases derived purely from theoretical calculations, methods for generating pharmacophores and the identification of potentially useful molecules using 3D pharmacophore searching. Some background material concerning molecular conformations and energy calculations can be found in Appendix 2.

27

2.   EXPERIMENTAL 3D DATABASES

The data stored in a 3D database either comes from experiment or can be calculated using a computational method. Of the experimental sources, the best known are those involving structures solved using x-ray crystallography. The Cambridge Structural Database (CSD) contains the x-ray structures of more than 400,000 organic and organometallic compounds [Allen et al. 1979; Allen 2002]. The Protein Data Bank (PDB) contains more than 44,000 x-ray and nuclear magnetic resonance (NMR) structures of proteins and protein–ligand complexes and some nucleic acid and carbohydrate structures [Bernstein et al. 1977; Berman et al. 2000]. Both these databases are widely used and continue to grow rapidly.

The CSD was established in 1965 and is maintained by the Cambridge Crystallographic Data Centre. It is possible within the CSD to retrieve structures using many types of textual and structural queries, including substructure searching. Thus one may retrieve the crystal structure(s) of an individual molecule (e.g. aspirin) or all structures containing a quinoline ring system. It is also possible to retrieve structures that meet specific 3D constraints. For example, one may wish to retrieve all crystal structures in which a specific ring system adopts a particular conformation (e.g. all examples of boat cyclohexane) or in which two functional groups are separated by a particular distance or distance range.

Whereas the CSD is a true database system, the PDB is more a communal repository of data files, one for each protein structure. Founded in 1971 the PDB now contains approximately 44,000 structures, most obtained using x-ray crystallography but with some determined using NMR. The PDB is most commonly accessed via a web interface, which enables structures to be retrieved using various textual queries (such as by author, protein name, literature citation). Some web interfaces also enable searches to be performed using amino acid sequence information. As the number of protein structures has grown so it has been recognised that the "flat file" system is inappropriate and more modern database systems and techniques have been introduced based on the information in the PDB. Several groups have also developed database systems that enable chemical and/or geometric features to be included. For example, it is possible in the Relibase system [Hendlich 1998] to search for protein–ligand complexes where the ligand contains a particular substructure, or where the ligand makes some specific intermolecular interaction with the protein.

The CSD and the PDB have also been used extensively for *data mining*, to derive knowledge and rules about conformational properties and intermolecular interactions. This knowledge can then be used in theoretical methods such as conformational search algorithms or protein–ligand docking programs.

A straightforward example is the analysis of the CSD to determine how bond lengths depend upon the nature and environment of the atoms involved [Allen et al. 1987]. The CSD is also invaluable for providing information about the conformations of specific molecular fragments. For example one may identify the low-energy conformations of a particular ring system or determine the preferred torsion angles about specific bonds, as illustrated in Figure 2-1.

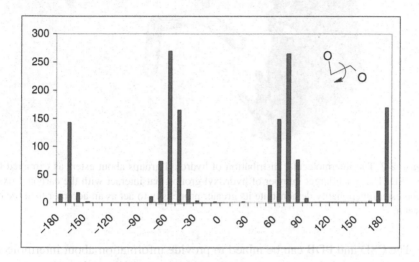

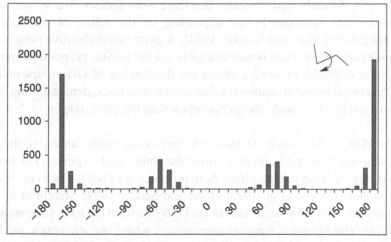

Figure 2-1. A comparison of the torsion angle distributions for the O–C–C–O fragment (top) and the C–C–C–C fragment (bottom), derived from the CSD. In each case all atoms were required to be sp^3 hybridised and only acyclic bonds were permitted. The distributions show the distinct preference of the OCCO fragment to adopt gauche torsion values (±60°) whereas CCCC favours anti-values (±180°).

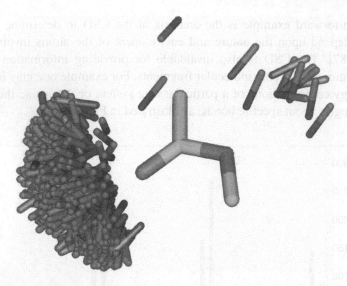

Figure 2-2. The intermolecular distribution of hydroxyl groups about esters as extracted from the CSD. The much larger number of hydroxyl groups that interact with the carbonyl oxygen in the ester functionality demonstrate its greater propensity to act as an acceptor than the ether oxygen.

The CSD and PDB can be mined to provide information about intermolecular interactions. An early study revealed that there were distinct distance and angular preferences for hydrogen bonds, depending on the nature of the donor and acceptor [Murray-Rust and Glusker 1984]. A more comprehensive range of such intermolecular interactions is now available via the IsoStar program [Bruno et al. 1997]. For example, Figure 2-2 shows the distribution of OH groups about the ester functional group in small-molecule crystal structures, demonstrating that the carbonyl oxygen is a much stronger acceptor than the ether oxygen.

The PDB has been used extensively to further our understanding of the nature of protein structure and its relationship to the amino acid sequence. For example, various classification schemes have been proposed for dividing protein structures into families [Orengo et al. 1993; Holm and Sander 1994; Murzin et al. 1995; Brenner et al. 1997]. The structures in the PDB also form the basis for *comparative modelling* (also known as *homology modelling*), where one attempts to predict the conformation of a protein of known sequence but unknown structure using the known 3D structure of a related protein. The PDB has also provided information about the nature of the interactions between amino acids, and between proteins and water molecules and small-molecule ligands.

3. 3D PHARMACOPHORES

A major use of 3D database systems is for the identification of compounds that possess 3D properties believed to be important for interaction with a particular biological target. These requirements can be expressed in a variety of ways, one of the most common being as a *3D pharmacophore*. A 3D pharmacophore is usually defined as a set of features together with their relative spatial orientation. Typical features include hydrogen bond donors and acceptors, positively and negatively charged groups, hydrophobic regions and aromatic rings. The use of such features is a natural extension of the concept of *bioisosterism*, which recognises that certain functional groups have similar biological, chemical and physical properties [Thornber 1979; Patani and LaVoie 1996]. Some common bioisosteres are shown in Figure 2-3.

3D pharmacophore searching can be particularly useful when trying to identify structures that have the desired activity but which are from a previously unexplored chemical series (sometimes referred to as "lead hopping") [Sheridan and Kearsley 2002]. This is due to the more abstract nature of the 3D pharmacophore; it depends on atomic properties rather than element types, and not on any specific chemical connectivity.

The spatial relationships between the features in a 3D pharmacophore can be specified as distances or distance ranges or by defining the (*xyz*) locations of the features together with some distance tolerance (typically as a spherical tolerance region). A simple example of a 3D pharmacophore based purely upon distance ranges is given in Figure 2-4. It may also be possible to include other geometric

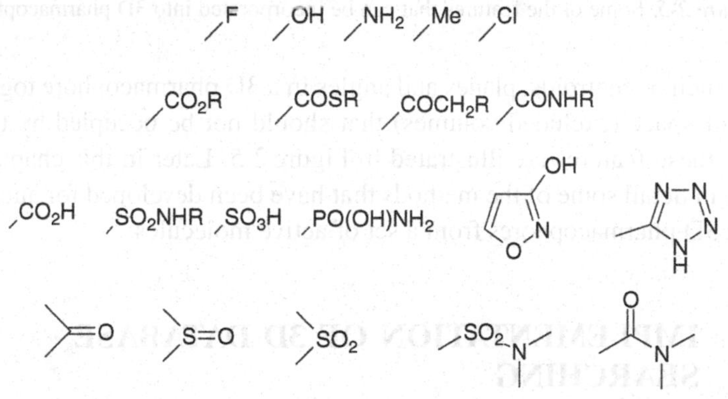

Figure 2-3. A selection of common bioisosteres; each line contains a set of distinct functional groups that can often be substituted for each other whilst retaining the biological activity.

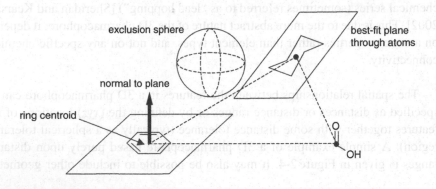

Figure 2-4. An example of a simple H₁ antihistamine pharmacophore together with some typical inhibitors [ter Laak et al. 1995].

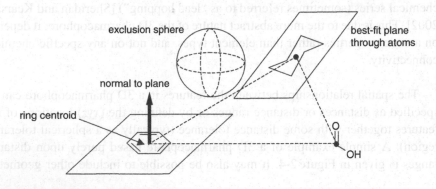

Figure 2-5. Some of the features that can be incorporated into 3D pharmacophores.

features such as centroids, planes and angles in a 3D pharmacophore together with regions of space (excluded volumes) that should not be occupied by the ligand. Some of these features are illustrated in Figure 2-5. Later in this chapter we will consider in detail some of the methods that have been developed for automatically deriving 3D pharmacophores from a set of active molecules.

4.   IMPLEMENTATION OF 3D DATABASE SEARCHING

In common with the substructure searching methods discussed in Chapter 1, a two-stage procedure is most commonly used to perform 3D searches of databases such as the CSD where the objective is to identify conformations that match the query. The first stage employs some form of rapid screen to eliminate molecules that could not match the query. The second stage uses a graph-matching procedure to identify those structures that do truly match the query.

One way to achieve the initial screen is to encode information about the distances between relevant groups in the molecular conformation. This can be achieved using bitstrings in which each bit position corresponds to a distance range between a specific pair of atoms or groups of atoms (e.g. functional groups or features such as rings). For example, the first position may correspond to the distance range 2–3Å between a carbonyl oxygen and an amine, the second position to the range 3–4Å between these two features and so on. Initially the bitstring (or *distance key*) would contain all zeros. For each molecular conformation, the distances between all pairs of atoms or groups are calculated and the appropriate bits set to "1" in the distance key. It is common to use smaller distance ranges for the more common distances found in 3D structures and larger distance ranges for the less common distances; this provides a more even population distribution across the distance bins than were a uniform binning scheme employed [Cringean et al. 1990]. In addition to pairwise distances it is also possible to encode angular information involving sets of three atoms and torsional information from four atoms to further improve the search efficiency [Poirrette et al. 1991, 1993].

To perform a search, the distance key for the query is generated and used to screen the database. Only those structures that remain after the screening process are subjected to the more time-consuming second stage. This typically involves a subgraph isomorphism method such as the Ullmann algorithm. The key difference between the 2D and 3D searches is that in the former the only edges in the molecular graph are those between atoms that are formally bonded together. By contrast, in the corresponding 3D graph there is an edge between all pairs of atoms in the molecule; the value associated with each edge is the appropriate interatomic distance. Thus a 3D graph represents the *topography* of a molecule. The query is often specified in terms of distance ranges; should the distance in the database conformation fall within the bounds of these ranges then that would be counted as a potential match. Potential matches are then confirmed by fitting the relevant conformation to the query in Cartesian space. This final fitting step is needed, for example, because stereoisomers cannot be distinguished using distance information alone.

5.   THEORETICAL 3D DATABASES

Experimental databases such as the CSD are extremely valuable sources of information, but for most compounds no crystal structure is available. There is also an increasing need to evaluate virtual compounds – structures that could be synthesised (e.g. using combinatorial chemistry) but which have not yet been made. Even when experimental data is available, this usually provides just a single conformation of the molecule which will not necessarily correspond to the active conformation; most molecules have a number of conformations accessible to them. It is thus desirable to include mechanisms for taking the conformational space of the molecules into account during 3D database searching.

5.1   Structure-Generation Programs

One of the main reasons for the surge of interest in 3D database searching in the 1990s was the development of *structure-generation program* that can take as input a 2D representation of a molecule such as a connection table or a SMILES string and generate a single, low-energy conformation. The two most widely used structure-generation programs are CONCORD [Rusinko et al. 1988] and CORINA [Gasteiger et al. 1990]. Both of these methods use a broadly similar approach. The first step is to identify features such as rings, bond orders and stereocentres in the molecule. Of particular interest for structure generation are rings and ring systems; whereas it is relatively straightforward to construct a sensible 3D structure for the acyclic portions of the molecule from standard bond lengths, bond angles and torsional preferences, this is more difficult for the cyclic parts. Typically, the rings and ring systems are constructed separately using a library of predefined structures and a set of rules that determines how they should be connected together. Next, the acyclic side chains are added in a default low-energy conformation and finally the structure is adjusted to deal with any high-energy, unfavourable interactions.

Most structure-generation programs are designed to produce a single conformation for each molecule. The earliest 3D database systems thus contained just one conformation for each molecule. Whilst this represented a significant advance, the limitations of using a single conformation were widely recognised. One way to deal with this problem is to modify the conformation at search time. This is usually achieved by "tweaking" the conformation. Given a set of distance constraints between pairs of atoms it is possible to calculate the values that the torsion angles of the intervening rotatable bonds should adopt in order to enable the constraints to be satisfied [Shenkin et al. 1987; Hurst 1994]. This therefore provides a mechanism for generating a conformation that matches the pharmacophore.

An alternative to the tweak method is to store multiple conformations for each molecule in the database. A large number of methods have been introduced for exploring conformational space; these methods are crucial for a number of procedures in chemoinformatics and computational chemistry. In the next section we therefore provide a summary of some of the key methods with a particular focus on the two that are most commonly used: systematic search and random search.

5.2   Conformational Search and Analysis

The usual goal of a *conformational analysis* is the identification of all accessible minimum-energy structures of a molecule. Even for simple molecules there may be a large number of such structures, making conformational analysis a difficult problem. The minimum with the lowest energy is referred to as the *global minimum-energy conformation*. It is important to note that the global minimum-energy conformation may not necessarily correspond to a biologically active 3D structure.

The usual strategy in conformational analysis is to use a search algorithm to generate a series of initial conformations. Each of these in turn is then subjected to energy minimisation (see Appendix 2) in order to derive the associated minimum-energy structure [Leach 1991, 2001; Scheraga 1993]. Molecular mechanics is usually used for the energy calculation and minimisation step; only for rather small systems is it feasible to consider quantum mechanical methods.

5.3   Systematic Conformational Search

In the *systematic search method* conformations are generated by systematically assigning values to the torsion angles of the rotatable bonds in the molecule. Each torsion angle is restricted to a predetermined set of possible values. The simplest type of systematic search algorithm is the *grid search*, in which conformations corresponding to all possible combinations of torsion angle values are generated. As a simple example, suppose we have a molecule with two variable torsion angles in which the first torsion angle adopts the values 60°, 180° and −60° and the second torsion angle adopts values 0° and 180°. A total of six conformations will be generated by the grid search (i.e. 60°, 0°; 60°, 180°; 180°, 0°; 180°, 180°; −60°, 0°; −60°; 180°), each of which is then subjected to energy minimisation.

If the number of values permitted to torsion angle i is n_i and there are N variable torsion angles in the molecule then the total number of conformations C generated by the grid search is:

$$C = \prod_{i=1}^{N} n_i \tag{2.1}$$

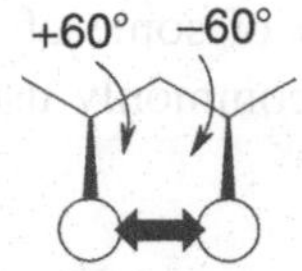

+60°   −60°

Figure 2-6. The gauche interaction arises from two adjacent bonds in an alkyl chain having torsion angles of +60 and −60.

The number of structures thus grows exponentially with the number of rotatable bonds in the molecule. This is sometimes referred to as a *combinatorial explosion*. However, significant improvements in efficiency can be made to the grid search if one recognises that a large proportion of the structures so generated often contain high-energy, unfavourable interactions due to clashes between parts of the structure. It is inappropriate to subject these structures to the time-consuming energy minimisation step. These unfavourable interactions often arise from particular combinations of torsion angles. A simple example is the so-called *gauche interaction*, which results from two adjacent bonds in an alkyl chain having torsion angle values of 60° and −60° (Figure 2-6).

Having identified a problem such as this we would like to eliminate from further consideration any other structure that contains the same offending combination of torsion angles. This can be achieved using a *depth-first search with tree pruning*. We first determine the order in which the torsion angles will be varied. We then try to generate a conformation in which all torsion angles have their first value assigned. If this is successful then the second conformation is generated by modifying the value of the last torsion angle; when all of its values are exhausted the algorithm *backtracks* to consider the penultimate torsion angle, and so on. This can be represented in a tree-like structure, as shown in Figure 2-7. Should a problem be encountered the algorithm immediately rejects all combinations that lie below that particular node in the tree.

It is important when using this method that the torsion angles are considered in an appropriate sequence so that once the conformation of a part of the molecule has been assigned it will not be changed by any subsequent torsional assignments. Structures containing rings can be difficult for the systematic search method. It is necessary to break an appropriate number of ring bonds to give a pseudo-acyclic structure which is then subjected to the search as usual. It is also necessary to include extra checks on the resulting conformation to ensure that the ring is formed properly.

When all of the nodes in the search tree have been considered the search is complete; the defined end point is one of the advantages of the systematic search

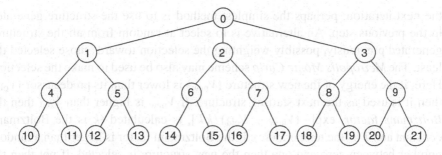

Figure 2-7. Tree representation of conformational space used in the depth-first search with backtracking. In this simple example there are three variable torsion angles with three, two and two values permitted to each respectively, giving a total of 12 possible conformations. Each node in the tree represents a state in which between one and three of these torsions have been assigned. The nodes are visited in the sequence 0–1–4–10–4–11–4–1–5–12–5–13–5–1–0–2–6–14 and so on. Should a problem be detected the algorithm immediately backtracks to the appropriate node. For example, if a problem arises at node 4 then the algorithm moves to node 5; a problem at node 2 leads to node 3.

when compared to most other methods. Several extensions that aim to improve the efficiency and utility of the basic systematic search have been described [Bruccoleri and Karplus 1987; Lipton and Still 1988; Goodman and Still 1991]. One strategy to deal with the combinatorial explosion is to use larger fragments, or building blocks [Gibson and Scheraga 1987; Leach et al. 1987, 1988]. This is because there are usually many fewer combinations of fragment structures than combinations of torsion angles. Fragment building is particularly useful when the molecule contains rings and ring systems, which, as noted above, cause particular difficulties for the systematic search method.

5.4   Random Conformational Search

A number of random conformational search methods have been devised [Saunders 1987; Li and Scheraga 1987; Ferguson and Raber 1989; Chang et al. 1989]. All of them involve some form of iterative procedure in which a structure is selected from those previously generated, randomly modified and then minimised. If this results in a new conformation it is added to the list of structures found. The process is then repeated. In contrast to the systematic search methods there is no natural end point for a random search; the process continues until either a predefined number of iterations has been attempted and/or until no new conformations can be generated. Different random search methods differ in the way that the structure is modified and in the way that the structure is selected for the next iteration. Modifications are most frequently achieved by either varying torsion angles (keeping the bond lengths and angles fixed) or by changing the (*xyz*) coordinates of the atoms. There are many ways to select the structure for the next iteration; perhaps the simplest method is to use the structure generated in the previous step. An alternative is to select at random from all the structures generated previously, possibly weighting the selection towards those selected the least. The *Metropolis Monte Carlo* scheme may also be used to make the selection. Here, if the energy of the new structure (V_{new}) is lower than its predecessor (V_{old}) then it is used as the next starting structure. If V_{new} is higher than V_{old} then the *Boltzmann factor*, $\exp[-(V_{new} - V_{old})/kT]$, is calculated ($k$ is the Boltzmann constant and T is the temperature). If the Boltzmann factor is larger than a random number between zero and one then the new structure is selected. If not then the previous structure is retained. The Metropolis method therefore makes it possible for higher-energy structures to be selected; these may correspond to previously unexplored regions of conformational space.

A commonly used version of the Monte Carlo method is *simulated annealing* [Kirkpatrick et al. 1983]. Here, the temperature is gradually reduced from a high value to a low value. At high temperatures the system can overcome high-energy barriers due to the presence of the temperature in the denominator of the Boltzmann factor; this enables it to explore the search space very widely. As the temperature falls so the lower energy states become more probable. By analogy with the physical process of annealing, used in the manufacture of very large single crystals of materials such as silicon, at absolute zero the computational system should exist in the global minimum-energy state. In the case of conformational analysis this would correspond to the global minimum-energy conformation. However, this would require an infinite number of temperature decrements at each of which the system would have to come to equilibrium. Thus in practice it is not guaranteed to find the global energy minimum. Several simulated annealing runs would therefore typically be performed.

5.5   Other Approaches to Conformational Search

There are many alternatives to the systematic and random methods; two that have been quite widely used are distance geometry and molecular dynamics. *Distance geometry* [Crippen 1981; Crippen and Havel 1988; Blaney and Dixon 1994] uses a description of molecular conformation based on interatomic distances allied to various mathematical procedures to generate structures for energy minimisation. *Molecular dynamics* solves Newton's equations of motion for the atoms in the system, to give a *trajectory* that defines how the positions of the atoms vary with time. Snapshots taken from the trajectory may then be subjected to minimisation in order to generate a sequence of minimum-energy conformations.

The choice of conformational search method is governed by many factors; some methods in particular require a significant amount of user input to set up each calculation and are therefore more suited to detailed analyses of individual molecules. A common goal in such cases is to identify the bioactive conformation, for example of a cyclic peptide. Other techniques are more amenable to automation, and so can be used for the construction of 3D databases and various kinds of virtual screening calculations on large collections of molecules. Such calculations can involve very large numbers (millions) of molecules and so it is important that the conformational search method is automatic (i.e. able to operate without user intervention), efficient and robust.

5.6 Comparison and Evaluation of Conformational Search Methods

Historically, conformational search methods were usually compared by examining a very small number of rather complex molecules. A classic example was the comparison of five different search methods (systematic search, Cartesian and torsional random methods, distance geometry and molecular dynamics) using cycloheptadecane ($C_{17}H_{34}$) [Saunders et al. 1990]. Recent studies have concentrated on comparison with experiment, using databases of structures solved by x-ray crystallography. Structures from the PDB have been of particular interest as the ligand conformations correspond to the presumed bioactive conformations of the molecules. Although the PDB contains a large number of protein–ligand complexes, not all of these may be considered suitable for a comparative study. It is common to restrict the ligands to small, "drug-like" molecules taken from high-resolution structures. An example of such a study is that of Boström (2001) who evaluated five different programs against a set of 32 ligands. Various performance criteria were used, including the ability of each program to generate the bioactive conformation together with the number of conformations generated by each method and the speed of calculation. Also of interest are the energetic changes that accompany ligand binding and whether the bound conformation is at the global or a local energy minimum. The strain energy that accompanies binding is found to vary somewhat; the study of Boström et al. [1998] suggested that for approximately 70% of the systems they studied the bioactive conformation was within 3 kcal/mol of the global energy minimum whereas Perola and Charifson [2004] found that whilst 60% of ligands bound with strain energies less than 5 kcal/mol at least 10% of ligands bound with strain energies greater than 9 kcal/mol. This latter study also highlighted the need to use higher-energy thresholds for more flexible molecules to ensure that the bioactive conformation was included in the ensemble of conformations generated.

5.7   The Generation of Distance Keys for Flexible Molecules

The database systems currently used for 3D pharmacophore searching enable very complex queries to be formulated, involving the use of many kinds of pharmacophoric and geometric features. It is also possible to combine both 2D and 3D information into the same query. Nevertheless, searches are still generally performed using the principles outlined earlier, with an initial screen based on a distance key being followed by a more time-consuming exact match.

When a set of conformations is available from experiment or from a computational method then it is usual to combine the distance keys for all the conformations of the molecule. This is done by taking the union of the distance keys for the individual conformations to give a single *ensemble distance key* that encodes the accessible conformational space of the molecule. For those systems that store just a single conformation and use a tweak-based approach it is still possible to determine constraints on the various interfeature distances from the 2D structure. One approach that has been employed is *triangle smoothing*. This procedure relies on the fact that combinations of three distances must meet certain geometrical requirements. There are two triangle inequality rules. First, the maximum value of the distance *AC* cannot be more than the sum of the maximum values of the distances *AB* and *BC*. Second, the minimum value of the distance *AC* cannot be less than the difference between the minimum value of *AB* and the maximum value of *BC* (see Figure 2-8). By repeatedly applying these constraints to the distance ranges within a molecule it is possible to derive lower and upper bounds on the interatomic and inter-feature distances.

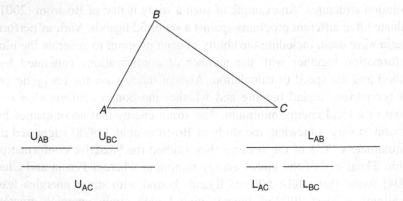

Figure 2-8. Graphical illustration of the two triangle inequalities: $U_{AC} \leq U_{AB} + U_{BC}$; $L_{AC} \geq L_{AB} - U_{BC}$ where L indicates the lower bound and U the upper bound.

6. METHODS TO DERIVE 3D PHARMACOPHORES

How can 3D pharmacophores be determined? Traditionally, 3D pharmacophore approaches are used when active molecules are known but where the structure of the target receptor is not available. The process of deriving a 3D pharmacophore is called *pharmacophore mapping*. There are two key issues to consider when deriving 3D pharmacophores. The first issue concerns the conformational flexibility of the molecules and how to take this into account. The second problem concerns the fact that there may be many different combinations of pharmacophoric groups within the molecules. As a consequence, there may be hundreds of potential 3D pharmacophores. The objective is to determine which of these potential pharmacophores best fits the data. In general, the aim is to identify the 3D pharmacophore(s) that contains the largest number of features common to all of the active molecules, and where these common features can be presented by each molecule in a low-energy conformation. Note that the generation and use of 3D pharmacophores is therefore based upon the assumption that all of the molecules bind in a common manner to the biological target.

Various methods have been devised for performing pharmacophore mapping; in the following sections we will discuss four of the most commonly used techniques: constrained systematic search, clique detection, the maximum likelihood method and the GA approach.

6.1 Pharmacophore Mapping Using Constrained Systematic Search

The *constrained systematic search* approach has its origins in the development of efficient algorithms for systematically exploring conformational space [Motoc et al. 1986]. As explained in the discussion on systematic conformational search methods, such procedures have an exponential dependence on the number of rotatable bonds in the molecule. By making use of tree-pruning methods (described earlier) to eliminate unfavourable, high-energy conformations the efficiency of these algorithms can be considerably improved. The constrained systematic search procedure for pharmacophore generation further improves the efficiency by deriving additional constraints on the torsional angles.

The first part of the procedure is to identify the pharmacophoric groups in each molecule that will be overlaid in the final pharmacophore. The most rigid molecule is then taken and its conformational space explored. During the conformational search the distances between all pairs of the selected pharmacophoric groups are recorded. The second most rigid molecule is then taken, and using the

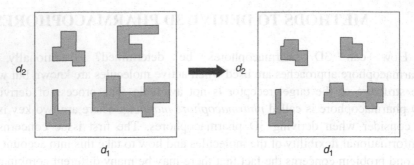

Figure 2-9. As successive molecules are considered in the constrained systematic search procedure so the distance constraints between pharmacophoric features become more restricted. Thus the figure on the left shows the distances permitted to the first molecule; these regions may be further restricted after analysis of the second molecule. (Adapted from Leach 2001.)

inter-pharmacophore distance ranges derived from the first molecule, constraints on the values permitted to each of its torsion angles are derived. Thus the only torsion angle values explored for the rotatable bonds in the second molecule are those that may permit it to match the pharmacophore distances found for the first molecule. As more and more molecules are considered so the distance ranges become more and more restricted (see Figure 2-9). When the more flexible compounds are considered the expectation is that only very limited ranges are possible on each of its torsion angles, making the search very efficient.

The classic application of the constrained systematic search procedure is to the discovery of 3D pharmacophores for the inhibition of Angiotensin Converting Enzyme (ACE), an enzyme involved in the regulation of blood pressure [Dammkoehler et al. 1989]. Typical inhibitors of ACE, some of which are shown in Figure 2-10, contain a zinc-binding group such as a thiol or a carboxylic acid, a terminal carboxylic acid group that interacts with a basic residue in the protein and an amido carbonyl group which acts as a hydrogen bond acceptor. The constrained systematic search method derives distance ranges between these features, accessible to all of the ACE inhibitors in a low-energy conformation. Two pharmacophores result from this analysis. Note that one of the pharmacophoric features is an *extended point*, corresponding to the presumed location of the zinc atom in the protein. The constrained systematic search was three orders of magnitude faster than an equivalent systematic search on each individual molecule, illustrating the advantage of using prior information to restrict the conformational search space.

A drawback of the constrained systematic search approach is that it is necessary to specify manually beforehand which pharmacophoric groups in each molecule are involved in interactions with the receptor, and the correspondences

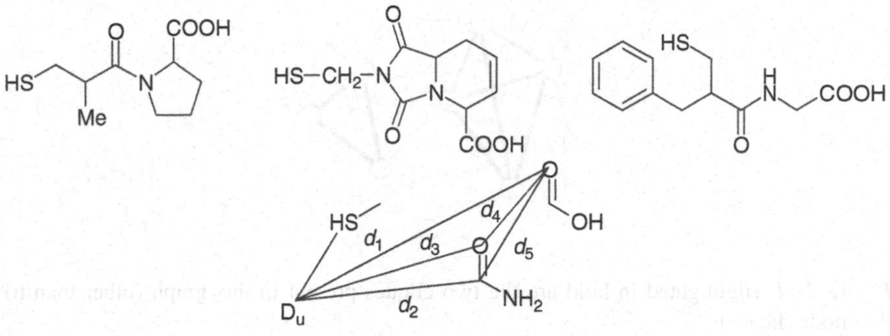

Figure 2-10. Some typical ACE inhibitors together with the features and the five distances used to define the 3D pharmacophores as discovered by the constrained systematic search method [Dammkoehler et al. 1989]. The two pharmacophores found correspond to different values of the distances $d_1 - d_5$. Du indicates the presumed location of the zinc atom in the enzyme (the extended point).

between these groups. It may be difficult to do this when there are many potential pharmacophores in the molecules, giving rise to many possible matches. The three remaining methods that we will discuss do not require the user to specify such correspondences; rather the algorithms determine them automatically.

6.2 Pharmacophore Mapping Using Clique Detection

A *clique* in a graph is a *completely connected subgraph*. It contains a subset of the nodes in the graph such that there is an edge between every pair of nodes in the subgraph. A key characteristic of a clique is that it cannot be extended by the addition of extra nodes; a simple example is shown in Figure 2-11. A *maximum clique* is the largest such subgraph that is present. An efficient algorithm for clique detection has been developed by Bron and Kerbosch (1973) and is often found to be a suitable approach for applications in chemoinformatics.

The clique detection approach to pharmacophore mapping can be explained using the following simple example. Consider one conformation for each of two molecules, each of which contains two donors (D_1, D_2 in the first molecule and d_1, d_2 in the second) and two acceptors (A_1, A_2 and a_1, a_2). These sets of features are illustrated in Figure 2-12. All possible pairs of matching features are generated. There are eight in this case: A_1a_1, A_1a_2, A_2a_1, A_2a_2, D_1d_1, D_1d_2, D_2d_1, D_2d_2. Each of these pairings constitutes a node in a new graph which is sometimes called a *correspondence graph*. An edge is created between all pairs of nodes in the correspondence graph where the corresponding inter-feature distances in the two molecules are equal (within some tolerance). If we assume for the sake of simplicity that the distances in our illustration need to be exactly equal then we

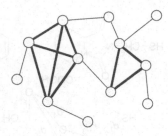

Figure 2-11. Highlighted in bold are the two cliques present in this graph (other than trivial two-node cliques).

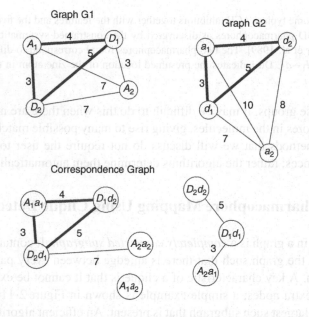

Figure 2-12. Simple illustration of clique detection method. Consider two molecules, each of which contains two donors and two acceptors. The two graphs at the top illustrate the distances between these sets of features in the two conformations being considered. There are eight nodes in the correspondence graph (bottom) but only six edges. From these, there is one maximum clique which consists of three pairs of matching features (bottom left). (Reproduced with permission from [Gillet and Leach 2006].)

can for example draw an edge between the nodes A_1a_1 and D_1d_2 (distance $= 4$) and also between the nodes D_2d_1 and A_2a_2 (distance $= 7$). The bottom half of Figure 2-12 shows the eight nodes of the correspondence graph with edges drawn between the matching nodes. This graph differs somewhat from those we have encountered previously, insofar that there is not necessarily a path between all nodes (it is a *disconnected graph*). In this case the two-node subgraph (A_1a_1, D_1d_2) does not constitute a valid clique as a third node (corresponding to the pairing D_2d_1) can be added to form the three-node clique shown in bold. This is the maximum clique present and represents a possible 3D pharmacophore consisting of three features. The corresponding subgraphs in the original graphs are shown in bold. The subgraph consisting of nodes (A_2a_1, D_1d_1 and D_2d_2) is not fully connected and therefore is not a clique, although it does contain two two-node cliques.

The first stage of the DISCO program which uses clique detection for pharmacophore mapping [Martin et al. 1993; Martin 2000; DISCO] involves the generation of a series of low-energy conformations for each molecule. The molecule with the smallest number of conformations is chosen as the "reference" molecule. Each of its conformations is considered in turn as the reference conformation. All the conformations of the other molecules in the set are then compared to this reference conformation and the cliques identified. Having considered each of the conformations of the reference molecule the entire set of cliques is then examined. Any clique that is common to all of the molecules, such that it is matched by at least one conformation of each molecule in the set, is a common pharmacophore.

6.3   The Maximum Likelihood Method for Pharmacophore Mapping

The *maximum likelihood method* Catalyst/HipHop [Barnum et al. 1996; Catalyst] also uses a pre-calculated set of low-energy conformations. Typically, these are obtained using *poling*, a conformational search method designed to generate a relatively small set of conformations that "covers" pharmacophore space [Smellie et al. 1995a, b]. The poling method adds an additional penalty term to the energy function during the minimisation part of the conformational analysis. This penalty term has the effect of "pushing" the conformation away from those found previously.

Having generated a set of conformations for each molecule, the first step in the maximum likelihood method is to identify all configurations of pharmacophoric groups that are present in the molecules. Each molecule is taken in turn as a reference structure and its conformations examined. All possible combinations of pharmacophoric groups contained within it are generated exhaustively (e.g. donor–acceptor–aromatic ring centroid; acceptor–acceptor–hydrophobe–donor). Each of these configurations of pharmacophoric groups in 3D space is then compared to the other molecules in the set in order to determine whether they possess a conformation that can be successfully superimposed on the configuration. In this step it is not required that all molecules match all of the features (i.e. a molecule can be active despite lacking a feature that is present in the binding motif of other active molecules).

This first step can generate a large number of possible configurations which are then scored and ranked according to how well the molecules map onto them and also according to their "rarity". The general strategy is to score more highly those configurations (referred to as *hypotheses*) that are well matched by the active molecules in the set but which are less likely to be matched by a large set of arbitrary molecules. The scoring function used is:

$$\text{score} = M \sum_{x=0}^{K+1} q(x) \log_2 \left(\frac{q(x)}{p(x)} \right) \tag{2.2}$$

where M is the number of active molecules in the set and K is the number of pharmacophoric groups in the hypothesis. The summation in Equation 2.2 takes account not only of the hypothesis itself but also other configurations containing $K-1$ features (there are K such configurations). An active molecule is assigned to the class $x = K + 1$ if it matches all the features, to one of the classes $x = 1$ through $x = K$ if it matches one of the configurations with one feature removed, or to the class $x = 0$ if it matches neither the full hypothesis nor one of the sub-configurations. $q(x)$ is the fraction of active molecules that matches each of the classes x. $p(x)$ corresponds to the fraction of a large set of arbitrary molecules that would match the class x configuration (i.e. the "rarity" value mentioned above). Pharmacophores that contain relatively uncommon features such as positively ionisable groups are less likely to be matched by chance than pharmacophores that contain more common features such as hydrophobic features and so have a smaller value of $p(x)$. In addition, the greater the geometrical distribution of features (dependent on the sum of the squares of the distances between the features and the common centroid) the higher the rarity value. A higher score for an hypothesis is thus associated with a higher value of $q(x)$ (i.e. more of the active molecules match) and lower values of $p(x)$ (i.e. it is less likely that an arbitrary, non-active molecule would match). The values of $p(x)$ are obtained using a mathematical model previously derived from an analysis of a set of biologically active molecules.

Another feature of the maximum likelihood method that contrasts with the constrained systematic search and the clique detection methods is its use of *location constraints* to define the pharmacophore rather than distance ranges between the features [Greene et al. 1994]. These location constraints usually correspond to a spherical region in 3D space, centred on a point, within which the relevant pharmacophoric feature should be positioned. The radius of the spherical region may vary; the different values reflect differences in the dependence of interaction energy on distance for different types of interaction. For example, an ionic interaction is typically more sensitive to distance changes than a hydrogen bond and so the tolerance on a charged feature would be correspondingly smaller. Another requirement is that the pharmacophoric features must be "accessible", rather than being buried inside the molecule.

6.4   Pharmacophore Mapping Using a Genetic Algorithm

Genetic algorithms (GAs) are a class of optimisation methods that are based on various computational models of Darwinian evolution [Goldberg 1989] (the term *evolutionary algorithms* is also used to describe such methods). GAs have found widespread use in many fields including computational chemistry and chemoinformatics [Clark and Westhead 1996; Jones 1998; Clark 2000]; we will discuss applications in protein–ligand docking, the generation of Quantitative Structure–Activity Relationships (QSAR) models, and library design in later chapters. In this section we will introduce the concepts of genetic and evolutionary algorithms using the pharmacophore-mapping procedure (GASP) described by Jones et al. (1995a, 2000; GASP). It should be noted that this represents just one way in which a GA can be implemented; there are many variants on the GA described here.

Genetic and evolutionary algorithms involve the creation of a *population* of potential solutions that gradually evolves towards better solutions. This evolution is dependent upon, and assessed using, a *fitness function* which provides a mechanism to score and therefore rank different members of the population. Another common feature is the use of a *chromosome* to encode each member of the population. The chromosome may be stored in a variety of ways, most commonly as a linear bitstring (i.e. a sequence of "0"s and "1"s). However it is also possible to store the chromosome as a sequence of integers or real numbers. The chromosomes form the basis for the generation of new potential members of the population.

In GASP the chromosome consists of $2N$-1 strings, where N is the number of molecules. N binary strings are used to represent the torsion angle values in the molecules (and so each specifies one molecule's conformation, as illustrated in Figure 2-13). Each angle is represented by eight bits (one byte); this allows the representation of integers in the range 0–255. The actual value encoded is rescaled to give a real number between 0 and 2π to be used as an angle of rotation. N-1 integer strings are used to represent the way in which a base molecule (chosen as the molecule with the smallest number of pharmacophore features) maps onto each of the other molecules. These mappings are used to superimpose each molecule in the appropriate conformation onto the base molecule using a molecular-fitting procedure (see Appendix 2). When the molecules have been superimposed the fitness value is calculated. The fitness function is dependent on

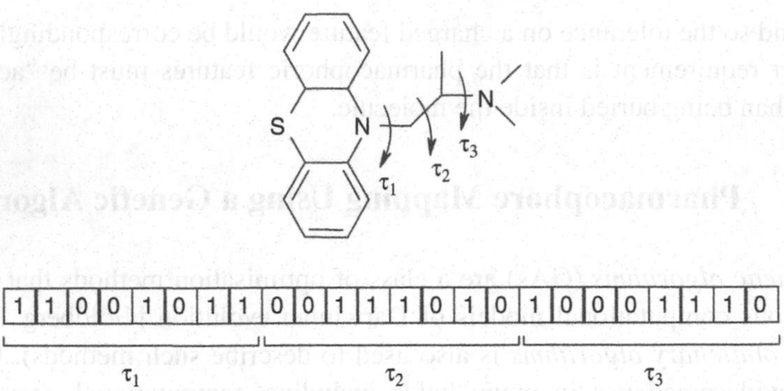

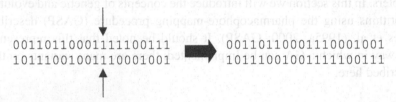

Figure 2-13. The three torsion angles of the molecule indicated are each encoded by one byte (eight bits) for use in the GA. Torsion angle τ_1 is $203(\equiv 1+2+8+64+128)$ which corresponds to an angle of rotation of 4.98 radians.

```
00110110001111100111              00110110001110001001
10111001001110001001              10111001001111100111
```

Figure 2-14. The crossover operator randomly selects a cross position and then exchanges the bits either side to give two new chromosomes.

the number and similarity of the features that are overlaid, the volume integral of the superimposed ensemble of conformations; and the van der Waals energy of the molecular conformations.

An initial population of solutions is first created via the random generation of a set of chromosomes. Each of these is scored using the fitness function. The population then evolves to explore the search space and (hopefully) to identify better solutions. To generate new members of the population the genetic operators *crossover* and *mutation* are applied. In crossover two parent chromosomes are taken, a cross position is randomly selected and two new child chromosomes are produced by swapping the bits either side of the cross position, as illustrated in Figure 2-14 for binary bitstrings. The mutation operator involves flipping a bit (for a binary chromosome) or altering the value of an integer to a new value chosen at random from the set of allowed values (for an integer string).

GASP uses *steady-state reproduction* in which a small number of children are created in each iteration; the same number of members are removed from the population and are replaced by the new children [Davis 1991]. Parent chromosomes are selected for reproduction using *roulette wheel selection* which biases the selection towards fitter members. It is analogous to spinning a roulette wheel with each member of the population having a slice of the wheel of size proportional to its fitness so that the "best" members of the population produce proportionately more offsprings. For example, if the fitness values are 2, 3, 5 and 6 then the relative sizes of the wheel slices will be 2/16, 3/16, 5/16 and 6/16 respectively. The crossover and mutation operators are then applied with the new children being scored and inserted into the population. Having completed one cycle of the GA, the procedure is repeated. The process continues either for a predetermined number of iterations or until convergence is reached (i.e. no better solution can be found). The output is the best-scoring solution found, according to the fitness function. A GA is at heart a random procedure (i.e. it is *non-deterministic*). Thus each run will typically generate a different solution and so it is usual to perform several runs to generate a series of solutions which are ranked according to their fitness values.

In GASP each molecule is initially represented by a single conformation and the conformational space is explored at search time. This contrasts with the previous approaches where multiple conformers are generated prior to pharmacophore generation and each conformer is subsequently treated as rigid molecule.

6.5 Other Approaches to Pharmacophore Mapping

The four pharmacophore mapping methods described above have been available for many years and some of their limitations have been highlighted [van Drie 2003]. For example, one study compared the performance of three of the methods using structures taken from protein–ligand complexes extracted from the PDB [Patel et al. 2002]. In this case the common pharmacophore could be unambiguously identified, but the results were disappointing. Recognising these limitations several groups have initiated the development of new approaches involving a range of interesting algorithms. For example, Feng et al. [2006] first represent the molecules in the set as bitstrings with one bitstring per molecular conformation. They use a modification of a Gibbs sampling algorithm that was previously applied to align DNA and protein sequences to identify a set of conformers, one for each molecule, that have specific bits in common. Clique detection is then used to define pharmacophore hypotheses based on the alignment. An interesting characteristic of the method is that it is able to find multiple hypotheses, which are each satisfied by different compounds thus allowing the identification of multiple binding modes. Richmond et al. [2004] have developed a novel method for the pairwise alignment of molecules that is based on an image recognition algorithm. Pairs of atoms, one from each molecule, are tentatively matched using an algorithm for solving the linear assignment problem. The initial atom-based and pairwise method has subsequently been modified to handle multiple molecules represented by pharmacophoric features to form a component of a new pharmacophore identification program. Recognising that a trade-off often exists between the alignment that can be achieved and the conformations adopted by molecules, Cottrell et al. [2004] have extended the GA approach to identify multiple plausible pharmacophore hypotheses. The benefits of these recent developments have yet to be proved, but given current efforts in this area, it is likely that significant progress will continue to be made.

6.6   Practical Aspects of Pharmacophore Mapping

Automated methods for pharmacophore mapping can offer significant benefits, but as with any other technique it is important to pay close attention to the construction of the set of molecules used as input and to the results produced. For most programs an "ideal" data set will contain a number of high-affinity molecules from several different chemical series that do not possess too much conformational freedom and do not have too many pharmacophoric features. These last two characteristics reduce the complexity of the search space. Stereoisomers or diastereoisomers of the same molecule that have different activities can be particularly informative when searching for a 3D pharmacophore. The output from a pharmacophore-mapping calculation should also be carefully examined to ensure that each active compound has a good geometric fit to the pharmacophore in a reasonable, low-energy conformation. Where appropriate the pharmacophore should also be consistent with any known relationships between the chemical structure and the biological activity. As we have discussed, the methods for pharmacophore mapping differ quite extensively and so it may be appropriate (if possible) to use more than one method for a given data set.

Several of the methods for pharmacophore mapping permit the use of *site points* (also known as *extended points*) in deriving the pharmacophore. These correspond to the presumed locations of atoms in the protein. The use of site points can account for the fact that two different ligand atoms may be able to bind to the same protein atom yet be positioned in quite different locations within the binding site, as illustrated in Figure 2-15. Indeed, it is possible to derive 3D pharmacophores directly from a protein x-ray structure by analysing the binding site and determining regions where ligand atoms or functional groups might be able to interact strongly with the protein. The active site of the protein may itself be represented through the use of *exclusion spheres* (see Figure 2-5). These are regions where a ligand atom may not be positioned (as it would clash with a protein atom). An alternative to the use of exclusion spheres is to define an *inclusion volume*; this is a region of space within which the ligand is required to fit.

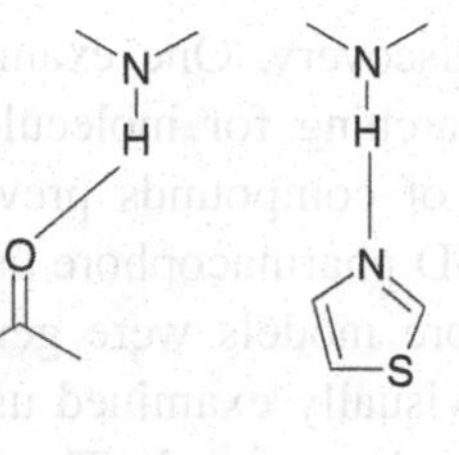

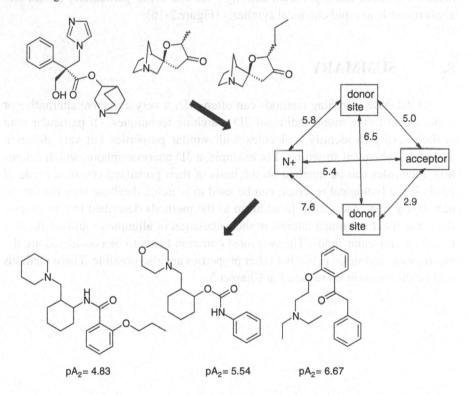

Figure 2-15. It is not always necessary for two atoms to be located in the same position in the binding site in order to interact with the same protein atom. In this case the carbonyl oxygen and the thiazole nitrogen both form a hydrogen bond with the same protein NH group but with different geometries.

pA₂= 4.83   pA₂= 5.54   pA₂= 6.67

Figure 2-16. Identification of muscarinic M_3 receptor antagonists using 3D pharmacophore mapping and database searching [Marriott et al. 1999]. pA_2 is a measure of biological activity with a larger number representing a more potent compound.

7. APPLICATIONS OF 3D PHARMACOPHORE MAPPING AND 3D DATABASE SEARCHING

Since the introduction of commercial software for performing pharmacophore mapping and 3D database searching there has been a steady stream of publications describing the application of these methods, typically to the discovery of new chemical series for drug discovery. One example is that reported by Marriott et al. [1999] who were searching for molecules active against the muscarinic M_3 receptor. Three series of compounds previously reported in the literature were chosen to derive the 3D pharmacophore using the clique-detection method. Five possible pharmacophore models were generated, each containing at least four features. These were visually examined using molecular graphics and two were selected for the 3D database search. These two pharmacophores contained a positively charged amine, a hydrogen bond acceptor and two hydrogen bond donor sites. The 3D database searches gave 172 molecules for testing; three of these compounds had significant activity with one being particularly potent and also amenable to rapid chemical synthesis (Figure 2-16).

8.   SUMMARY

3D database searching methods can often offer a very attractive alternative or complement to the more traditional 2D searching techniques. Of particular note is their ability to identify molecules with similar properties but very different underlying chemical structures. For example, a 3D pharmacophore, which defines how molecules can be overlaid on the basis of their presumed common mode of binding at a biological receptor, can be used to search a database with the aim of identifying new lead series. In addition to the methods described in this chapter there has also been much interest in the calculation of alignments and similarities based on molecular fields. The two most common fields to be considered are the electrostatic and steric fields, but other properties are also possible. These methods will be discussed in more detail in Chapter 5.

Chapter 3

MOLECULAR DESCRIPTORS

1. INTRODUCTION

The manipulation and analysis of chemical structural information is made possible through the use of *molecular descriptors*. These are numerical values that characterise properties of molecules. For example, they may represent the physicochemical properties of a molecule or they may be values that are derived by applying algorithmic techniques to the molecular structures. Many different molecular descriptors have been described and used for a wide variety of purposes. They vary in the complexity of the information they encode and in the time required to calculate them. In general, the computational requirements increase with the level of discrimination that is achieved. For example, the molecular weight does not convey much about a molecule's properties but it is very rapid to compute. By contrast, descriptors that are based on quantum mechanics may provide accurate representations of properties, but they are much more time-consuming to compute. Some descriptors have an experimental counterpart (e.g. the octanol–water partition coefficient), whereas others are purely algorithmic constructs (e.g. 2D fingerprints).

In this chapter we will describe commonly used molecular descriptors that can be applied to relatively large data sets. These include descriptors that represent properties of whole molecules such as log P and molar refractivity; descriptors that can be calculated from 2D graph representations of structures such as topological indices and 2D fingerprints; and descriptors such as pharmacophore keys that require 3D representations of structures. We conclude by considering how descriptors can be manipulated and combined together. Our emphasis will be on descriptors that are properties of the whole molecule, rather than of substituents. The latter type of descriptor was very important in the development of QSARs, and will be briefly covered in Chapter 4.

2.   DESCRIPTORS CALCULATED FROM THE 2D STRUCTURE

2.1   Simple Counts

Perhaps the simplest descriptors are based on simple counts of features such as hydrogen bond donors, hydrogen bond acceptors, ring systems (including aromatic rings), rotatable bonds and molecular weight. Many of these features can be defined as substructures or molecular fragments and so their frequency of occurrence can be readily calculated from a 2D connection table using the techniques developed for substructure search. For most applications, however, these descriptors are unlikely to offer sufficient discriminating power if used in isolation and so they are often combined with other descriptors.

2.2   Physicochemical Properties

Hydrophobicity is an important property in determining the activity and transport of drugs [Martin and DeWitte 1999, 2000]. For example, a molecule's hydrophobicity can affect how tightly it binds to a protein and its ability to pass through a cell membrane. Hydrophobicity is most commonly modelled using the logarithm of the partition coefficient between n-octanol and water ($\log P$). The experimental determination of $\log P$ can be difficult, particularly for zwitterionic and very lipophilic or polar compounds; data are currently available for approximately 30,000 compounds only [Mannhold and van der Waterbeemd 2001]. Of course, there are no data for compounds not yet synthesised. There is thus considerable interest in the development of methods for predicting hydrophobicity values.

The first approach to estimating $\log P$ was based on an additive scheme whereby the value for a compound with a substituent X is equal to the $\log P$ for the parent compound (typically chosen to be the molecule with the substituent being hydrogen) plus the appropriate substituent constant π_X [Fujita et al. 1964]. The substituent constants, π_X, were calculated from experimental data as follows:

$$\pi_X = \log P_X - \log P_H \tag{3.1}$$

While the approach worked well within a congeneric series of compounds, the π-values were found not to be additive across different series. For example, it is inappropriate to use π-values derived from a benzene parent on electron-deficient rings such as pyridine.

Many other methods for estimating log P have been proposed. Some of these are based on breaking the molecule into fragments. The partition coefficient for the molecule then equals the sum of fragment values plus a series of "correction factors" to account for interactions between the fragments such as intramolecular hydrogen bonding. This leads to the following equation, first proposed by Rekker [1977, 1992]:

$$\log P = \sum_{i=1}^{n} a_i f_i + \sum_{j=1}^{m} b_j F_j \qquad (3.2)$$

where there are a_i fragments of type i with f_i being the corresponding contribution and b_j occurrences of correction factor j with F_j being the correction factor. The most widely used program of this type is the ClogP program, developed by Leo and Hansch [1993]. This contains a relatively small set of fragment values obtained from accurately measured experimental log P data for a small set of simple molecules. Moreover, it uses a wholly automatic fragmentation scheme; molecules are fragmented by identifying *isolating carbons*. These are carbon atoms that are not doubly or triply bonded to a heteroatom. Such carbon atoms and their attached hydrogens are considered hydrophobic fragments with the remaining groups of atoms being the polar fragments.

Two ClogP calculations are illustrated in Figure 3-1 for benzyl bromide and *o*-methyl acetanilide. The former contains one aliphatic isolating carbon six isolating aromatic carbons and one bromide fragment. To these fragment contributions

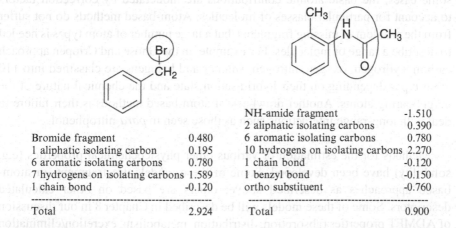

Bromide fragment	0.480	
1 aliphatic isolating carbon	0.195	
6 aromatic isolating carbons	0.780	
7 hydrogens on isolating carbons	1.589	
1 chain bond	-0.120	
Total	2.924	

NH-amide fragment	-1.510
2 aliphatic isolating carbons	0.390
6 aromatic isolating carbons	0.780
10 hydrogens on isolating carbons	2.270
1 chain bond	-0.120
1 benzyl bond	-0.150
ortho substituent	-0.760
Total	0.900

Figure 3-1. ClogP calculations on benzyl bromide and *o*-methyl acetanilide. (Adapted from Leach 2001.)

are added values for the seven hydrogens on the isolating carbons together with a contribution from one acyclic bond. *o*-methyl acetanilide contains an amide fragment, two aliphatic isolating carbons and six isolating aromatic carbons. There are also contributions from the hydrogen atoms, the acyclic bond, a benzyl bond (to the *o*-methyl group) and a factor due to the presence of an *ortho* substituent.

The advantage of fragment-based approaches is that significant electronic interactions can be taken into account; however, a potential disadvantage is that the methods fail on molecules containing fragments for which values have not been provided. This was a problem with early versions of ClogP where it was necessary to determine missing values using experimental measurements. However, the most recent version includes a method that estimates values for missing fragments [Leo and Hoekman 2000].

Atom-based methods provide an alternative to the fragment-based approaches. In these methods, log P is given by summing contributions from individual atoms [Ghose and Crippen 1986; Wang et al. 1997; Ghose et al. 1998; Wildman and Crippen 1999]:

$$\log P = \sum n_i a_i \tag{3.3}$$

where n_i is the number of atoms of atom type i and a_i is the atomic contribution. The atom contributions are determined from a regression analysis using a training set of compounds for which experimental partition coefficients have been determined. More details on regression methods are given in Chapter 4. In some cases, the basic atomic contributions are moderated by correction factors to account for particular classes of molecules. Atom-based methods do not suffer from the problem of missing fragments, but a large number of atom types is needed to describe a range of molecules. For example, in the Ghose and Crippen approach carbon, hydrogen, oxygen, nitrogen, sulphur and halogens are classified into 110 atom types depending on their hybridisation state and the chemical nature of the neighbouring atoms. Another drawback of atom-based methods is their failure to deal with long-range interactions such as those seen in *para*-nitrophenol.

Methods for the estimation of various other physicochemical properties (e.g. solubility) have been developed. Some of these models use fragment- or atom-based approaches as described above; others are based on other calculated descriptors. Some of these models will be described in Chapter 8 in our discussion of ADMET properties (absorption, distribution, metabolism, excretion/elimination and toxicity).

2.3   Molar Refractivity

The molar refractivity is given by:

$$MR = \frac{n^2 - 1}{n^2 + 2} \frac{MW}{d} \tag{3.4}$$

where n is the refractive index, d is the density, and MW is the molecular weight [Hansch and Leo 1995; Livingstone 2000]. Since molecular weight divided by density equals molar volume, molar refractivity is often used as a measure of the steric bulk of a molecule. The refractive index term accounts for polarisability of the molecule and does not vary much from one molecule to another. Molar refractivity can be calculated using atomic contributions in a similar way to the calculation of log P (the CMR program [Leo and Weininger 1995]).

2.4   Topological Indices

Topological indices are single-valued descriptors that can be calculated from the 2D graph representation of molecules [Lowell et al. 2001; Randić 2001]. They characterise structures according to their size, degree of branching and overall shape.

An early example is the Wiener Index [Wiener 1947] which involves counting the number of bonds between each pair of atoms and summing the distances, D_{ij}, between all such pairs:

$$W = \frac{1}{2} \sum_{i=1}^{N} \sum_{j=1}^{N} D_{ij} \tag{3.5}$$

The best-known topological indices are the molecular connectivity indices first described by Randić [1975] and subsequently developed by Kier and Hall [1986]. The *branching index* developed by Randić is calculated from the hydrogen-suppressed graph representation of a molecule and is based on the *degree* δ_i of each atom i. The degree equals the number of adjacent non-hydrogen atoms. A *bond connectivity value* is calculated for each bond as the reciprocal of the square root of the product of the degree of the two atoms in the bond. The branching index equals the sum of the bond connectivities over all of the bonds in the molecule:

$$\text{Branching Index} = \sum_{\text{bonds}} \frac{1}{\sqrt{\delta_i \delta_j}} \tag{3.6}$$

Kier and Hall generalised and extended the approach into a series of descriptors that became known as the *chi molecular connectivity indices*. First, the simple δ_i value is redefined in terms of the number of sigma electrons and the number of hydrogen atoms associated with an atom. Valence δ_i^v values are also introduced; these encode atomic and valence state electronic information through counts of sigma, pi and lone pair electrons. Thus, the *simple delta value* for atom i, δ_i is given by:

$$\delta_i = \sigma_i - h_i \tag{3.7}$$

where σ_i is the number of sigma electrons for atom i and h_i is the number of hydrogen atoms bonded to atom i. The *valence delta value* for atom i, δ_i^v is defined as:

$$\delta_i^v = Z_i^v - h_i \tag{3.8}$$

where Z_i^v is the total number of valence electrons (sigma, pi and lone pair) for atom i.

Table 3-1 shows the simple and valence delta values for various atoms. Thus, the simple delta value differentiates $-CH_3$ from $-CH_2-$. Whilst $-CH_3$ has the same simple delta value as $-NH_2$ it has a different valence delta value and so the two atoms can be differentiated using this descriptor. For elements beyond fluorine in the periodic table the valence delta expression is modified as follows:

$$\delta_i^v = \frac{Z_i^v - h_i}{Z_i - Z_i^v - 1} \tag{3.9}$$

where Z_i is the atomic number.

Table 3-1. Values of δ_i and δ_i^v for several common types of atom.

	$-CH_3$	$-CH_2-$	$=CH_2$	$-NH_2$
δ_i	1	2	1	1
δ_i^v	1	2	2	3

The chi molecular connectivity indices are sequential indices that sum the atomic delta values over bond paths of different lengths. Thus, the zeroth order chi index ($^0\chi$) is a summation over all atoms in a molecule (i.e. paths of length zero):

$$^0\chi = \sum_{\text{atoms}} \frac{1}{\sqrt{\delta_i}}; \quad ^0\chi^v = \sum_{\text{atoms}} \frac{1}{\sqrt{\delta_i^v}} \tag{3.10}$$

The first-order chi index involves a summation over bonds. The first-order chi index is thus the same as Randić's branching index when simple delta values are used:

$$^1\chi = \sum_{\text{bonds}} \frac{1}{\sqrt{\delta_i \delta_j}}; \quad ^1\chi^v = \sum_{\text{bonds}} \frac{1}{\sqrt{\delta_i^v \delta_j^v}} \tag{3.11}$$

Higher-order chi indices involve summations over sequences of two, three, etc. bonds.

To illustrate the differences between the various chi indices for a series of related structures we show in Figure 3-2 the values of the $^0\chi$, $^1\chi$ and $^2\chi$ indices for the isomers of hexane.

	Paths of length 2	Paths of length 3	Paths of length 4	Paths of length 5	$^0\chi$	$^1\chi$	$^2\chi$
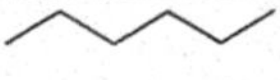	4	3	2	1	4.828	2.914	1.707
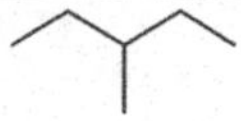	5	4	1	0	4.992	2.808	1.922
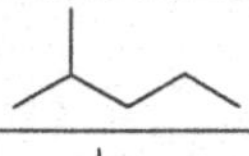	5	3	2	0	4.992	2.770	2.183
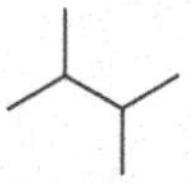	6	4	0	0	5.155	2.643	2.488
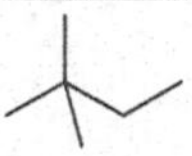	7	3	0	0	5.207	2.561	2.914

Figure 3-2. Chi indices for the various isomers of hexane. (Redrawn from Leach 2001.)

2.5   Kappa Shape Indices

The *kappa shape indices* [Hall and Kier 1991] are designed to characterise aspects of molecular shape by comparing a molecule with the "extreme shapes" that are possible for that number of atoms. As with the molecular connectivity indices, there are shape indices of various order (first, second, etc.). The first-order shape index involves a count over single bond fragments. The two extreme shapes are the linear molecule and the completely connected graph where every atom is connected to every other atom (Figure 3-3).

The first-order kappa index is defined as:

$$^1\kappa = \frac{2\, ^1P_{max}\, ^1P_{min}}{\left(^1P\right)^2} \tag{3.12}$$

where $^1P_{max}$ is the number of edges (or paths of length one) in the completely connected graph (see Figure 3-3); $^1P_{min}$ is the number of bonds in the linear molecule; and ^1P is the number of bonds in the molecule for which the shape index is being calculated. For a molecule containing A atoms, $^1P_{min}$ equals $(A-1)$ and $^1P_{max}$ is $A(A-1)/2$, since all pairs of atoms are connected. Thus $^1\kappa$ becomes:

$$^1\kappa = \frac{A\,(A-1)^2}{\left(^1P\right)^2} \tag{3.13}$$

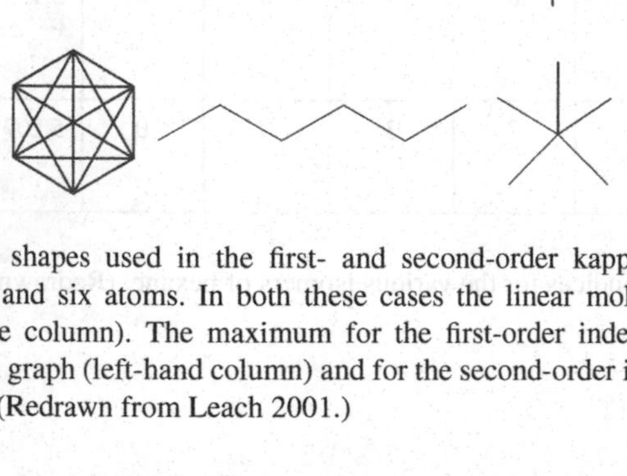

Figure 3-3. Extreme shapes used in the first- and second-order kappa indices for graphs containing four, five and six atoms. In both these cases the linear molecule corresponds to the minimum (middle column). The maximum for the first-order index corresponds to the completely connected graph (left-hand column) and for the second-order index to the star shape (right-hand column). (Redrawn from Leach 2001.)

The second-order kappa index is determined by the count of two-bond paths, written ^2P. The maximum value corresponds to a "star" shape in which all atoms but one are adjacent to the central atom ($^2P_{max} = (A-1)(A-2)/2$, Figure 3-3) and the minimum value again corresponds to the linear molecule ($^2P_{min} = A-2$). The second-order shape index is then:

$$^2\kappa = \frac{2^2 P_{max} \, ^2 P_{min}}{\left(^2 P\right)^2} = \frac{(A-1)(A-2)^2}{\left(^2 P\right)^2} \tag{3.14}$$

As with the molecular connectivity indices, higher-order shape indices have also been defined. The kappa indices themselves do not include any information about the identity of the atoms. This is achieved in the *kappa–alpha* indices where the alpha value for an atom i is a measure of its size relative to some standard (chosen to be the sp^3-hybridised carbon):

$$\alpha_i = \frac{r_i}{r_{Csp^3}} - 1 \tag{3.15}$$

Thus, the atom count A is adjusted for each non-sp^3-hybridised carbon (either by an increment or a decrement). The kappa–alpha indices are calculated as follows:

$$^1\kappa_\alpha = \frac{(A+\alpha)(A+\alpha-1)^2}{\left(^1 P + \alpha\right)^2} \tag{3.16}$$

$$^2\kappa_\alpha = \frac{(A+\alpha-1)(A+\alpha-2)^2}{\left(^2 P + \alpha\right)^2} \tag{3.17}$$

where α is the sum of the α_i's for all atoms in the molecule.

2.6 Electrotopological State Indices

The *electrotopological state indices* [Hall et al. 1991] are determined for each atom (including hydrogen atoms, if so desired) rather than for whole molecules. They depend on the *intrinsic state* of an atom, I_i, which for an atom i in the first row of the periodic table is given by:

$$I_i = \frac{\left(\delta_i^v + 1\right)}{\delta_i} \tag{3.18}$$

The intrinsic state thus encodes electronic and topological characteristics of atoms. The effects of interactions with the other atoms are incorporated by determining the number of bonds between the atom i and each of the other atoms, j. If this path length is r_{ij} then a perturbation ΔI_{ij} is defined as:

$$\Delta I_{ij} = \sum_j \frac{I_i - I_j}{r_{ij}^2}$$

(3.19)

The *electrotopological state* (*E-state*) for an atom is given by the sum of ΔI_{ij} and I_i. Atomic E-states can be combined into a whole-molecule descriptor by calculating the mean-square value over all atoms. Vector or bitstring representations can also be produced. There is a finite number of possible I values, and so a bitstring representation can be obtained by setting the appropriate bit for each of the different I values present in a molecule. Alternatively, a vector of real numbers can be derived by taking the sum or the mean of each unique intrinsic state.

Here, we have given details of some of the most widely used topological indices; many other indices have also been described and are available in software packages. For example, the widely used Molconn-Z program provides access to several hundred different descriptors [Molconn-Z].

2.7   2D Fingerprints

Some of the most commonly used descriptors are the 2D fingerprints described in Chapter 1. These descriptors were originally developed to provide a fast screening step in substructure searching. They have also proved very useful for similarity searching as will be described in Chapter 5. As explained in Chapter 1 there are two different types of 2D fingerprints: those based on the use of a fragment dictionary, and those based on hashed methods.

An advantage of dictionary-based fingerprints is that each bit position often corresponds to a specific substructural fragment. In some situations this can make interpretation of the results of an analysis more straightforward (e.g. if it is shown that specific bits are key to explaining the observed activity). This may have implications for the fragment dictionary: fragments that occur infrequently may be more likely to be useful than fragments which occur very frequently [Hodes 1976]. Unfortunately, the optimum set of fragments is often data set dependent. The hashed fingerprints are not dependent on a predefined dictionary so any fragment that is present in a molecule will be encoded. However, it is not possible to map from a bit position back to a unique substructural fragment and so the fingerprints are not directly interpretable. It is also important to realise that 2D fingerprints were originally designed to improve the performance of substructure searching algorithms rather than to be used as molecular descriptors. There is therefore no inherent reason why these descriptors should "work" for these alternative applications. The fact that they often do is probably due to the fact that a molecule's properties and biological activity often depends on features such as those encoded by 2D fingerprints. There have also been some attempts to redefine fragment dictionaries to make them more appropriate as molecular descriptors [Durant et al. 2002].

2.8   Atom Pairs and Topological Torsions

Atom pair descriptors encode all pairs of atoms in a molecule together with the length of the shortest bond-by-bond path between them [Carhart et al. 1985]. Each atom is described by its elemental type together with the number of non-hydrogen atoms to which it is bonded and its number of π-bonding electrons. An example of a simple atom pair is $CX2-(2)-CX2$, representing the $-CH_2-CH_2-$ sequence, Xn indicating the presence of n non-hydrogen neighbouring atoms. A more complex example is $CX2-(3)-O\cdot X1$, representing a fragment such as $A-CH_2-A=O$, A being any non-hydrogen atom and the dot symbol ("·") indicating the presence of a bonding π-electron. Atom pairs encode more distant information than is the case for the 2D fingerprints described above. The related *topological torsions* encode sequences of four connected atoms together with their types, number of non-hydrogen connections and number of π-electrons [Nilakantan et al. 1987]. It is straightforward to construct a bitstring descriptor in which each position represents the presence or absence of a particular atom pair or topological torsion in the molecule.

Kearsley et al. [1996] subsequently modified the atom pair definition in recognition of the fact that the properties of atoms are often more important than the specific element type. In their modified approach atoms are identified as belonging to one of seven binding property classes: cations, anions, neutral hydrogen bond donors and acceptors, atoms which are both donor and acceptor, hydrophobic atoms and all others.

Other descriptors based on similar ideas include, for example, the CATS (Chemically Advanced Template Search) descriptors [Schneider et al. 1999]. The Similog keys extend the approach to triplets of atoms and their topological distances with each atom being described by the presence or absence of the following four properties: donor, acceptor, bulkiness and electropositivity (known as the DABE atom typing scheme) [Schuffenhauer et al. 2003].

2.9   Extended Connectivity Fingerprints

Circular substructures were first developed for substructure searching in the DARC system where they were called FRELs (Fragment Reduced to an Environment that is Limited) [Attias 1983]. In FRELs the environment of an atom is described in terms of its neighbouring atoms up to a radius of two bonds. Similar descriptors called Extended Connectivity Fingerprints (ECFPs) and Functional Connectivity Fingerprints (FCFPs) [SciTegic] have been shown to be effective in similarity searching applications [Hert et al. 2004b]. The former are based on element types with the latter being based on generalised types. The extended connectivity of an atom is calculated using a modified version of the Morgan algorithm [Morgan 1965] and the degree of discrimination can be varied by extending the environment of the atom to include higher-order neighbours. An atom code is combined with the codes of its neighbours to the required level of description. For example, ECFP2 considers the first-order neighbours; ECFP4 extends to the third-order neighbours, etc.

2.10   BCUT Descriptors

BCUT descriptors were designed to encode atomic properties relevant to intermolecular interactions. They have been used extensively in diversity analyses. BCUT values are based on an earlier descriptor developed by Burden [1989] that is calculated from a matrix representation of a molecule's connection table. The atomic numbers of the non-hydrogen atoms are placed on the diagonal of the matrix. The off-diagonals are assigned the value 0.1 times the bond type if the atoms are bonded, and 0.001 if the atoms are not bonded. The two lowest eigenvalues (see Appendix 1) of this matrix are then calculated and combined to give a single index. The method was extended by Pearlman to generate a family of descriptors called the BCUT descriptors that could be used to define a low-dimensional chemistry space [Pearlman and Smith 1998]. Rather than using atomic number, the diagonal elements in the BCUT matrices encode atomic charge, atomic polarisability and atomic hydrogen bonding ability. The highest and lowest eigenvalues are then extracted for use as descriptors. The three types of matrix described above would thus give rise to six descriptors that describe a 6D space. It is also possible to derive BCUT values based on 3D properties. Here, the same atomic properties are encoded on the diagonals of the matrices with the off-diagonals encoding interatomic distances.

3. DESCRIPTORS BASED ON 3D REPRESENTATIONS

Given that molecular recognition depends on 3D structure, there has been considerable interest in the use of 3D descriptors. Such descriptors require the generation of 3D conformations; this can be computationally time-consuming when dealing with large data sets. Furthermore, a single low-energy conformation is usually insufficient and so conformational flexibility should be taken into account. It is usual to combine the descriptors for all of the conformations (sometimes weighted by the relative population of each conformation) to generate an ensemble descriptor that encodes the accessible conformational space of the molecule.

3.1   3D Fragment Screens

As described in the previous chapter, 3D screens were originally designed for use in 3D substructure searching. However, they can also be used as descriptors for applications such as library design and diversity analysis. These screens encode spatial relationships (e.g. distances and angles) between the different features of a molecule such as atoms, ring centroids and planes. Distance ranges for each pair of features are divided into a series of bins by specifying a bin width. For example, a distance range of 0–20 Å between two nitrogen atoms might be represented by ten bins, each of width 2 Å, covering the ranges 0–2.0 Å, 2.0–4.0 Å and so on. Alternatively, a non-uniform binning scheme may be used to provide a more even population distribution, as described in Chapter 2. Valence angle descriptors consist of three atoms, *ABC*, together with the angle between them (i.e. the angle between *AB* and *AC*). The atoms may or may not be bonded. Torsion angle descriptors consist of four atoms, *ABCD*, where *BC* is the torsion bond, and the torsion angle is the angle formed between *ABC* and *BCD*. As with valence angles, the atoms may be bonded or non-bonded. The different types of screens can be combined into a bitstring of length equal to the total number of bins over all feature types. Another distance-based descriptor uses the distances between triplets of atoms. Schemes for converting the three interatomic distances into a single value have also been devised [Bemis and Kuntz 1992; Nilakantan et al. 1993].

3.2   Pharmacophore Keys

Pharmacophore keys are an extension of the 3D screens developed for 3D pharmacophore searching [Pickett et al. 1996]. They are based on pharmacophoric features, that is, atoms or substructures that are thought to have relevance for receptor binding. Pharmacophoric features typically include hydrogen bond donors, hydrogen bond acceptors, charged centres, aromatic ring centres

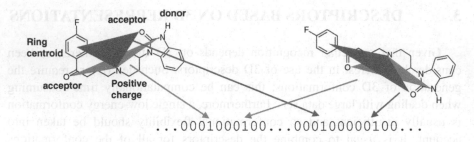

...0001000100...000100000100...

Figure 3-4. The generation of 3-point pharmacophore keys, illustrated using benperidol. Two different conformations are shown, together with two different combinations of three pharmacophore points. (Adapted from Leach 2001.)

and hydrophobic centres. Pharmacophore keys are most frequently based on combinations of either three or four pharmacophoric features. For example, a particular conformation of a molecule may contain a hydrogen bond acceptor 6 Å from an aromatic ring centroid and 5 Å from a basic nitrogen with the third distance being 7 Å. This 3-point pharmacophore is characterised by the types of pharmacophore features involved and the distances between them (Figure 3-4). In order to construct the binary pharmacophore key, all possible combinations of 3-point pharmacophores are enumerated together with all possible (binned) distances between the features. Each bit in the pharmacophore key bitstring thus represents one possible 3-point pharmacophore. The pharmacophore key for the molecule is then derived by setting the relevant bits in the key to "1" to indicate the 3-point pharmacophores it contains, considering all possible conformations of the molecule (Figure 3-4).

The first approach to the construction of pharmacophore keys used in the Chem-Diverse program [Davies 1996] was based on seven feature types and 32 distance ranges: <1.7 Å; 1.7–3.0 in 0.1 Å increments; 3.0–7.0 Å in 0.5 Å increments; 7–15 Å in 1 Å increments and >15 Å. There are 84 ways to select three features from seven and when these are combined with the 32 distance bins over 2 million combinations result. However, many of these pharmacophores are invalid since they violate the triangle inequality rule (see Chapter 2); moreover, some of the combinations are identical due to symmetry. When these are removed the total number of distinct 3-point pharmacophores reduces to approximately 890,000. 4-point pharmacophores contain more information (e.g. four points are required to differentiate stereoisomers) and can be more discriminating than the 3-point keys [Mason et al. 1999]. All combinations of four features together with 15 distance ranges give rise to over 350 million 4-point pharmacophores with valid geometry. Another extension is to combine information about the volume occupied by each conformation with the pharmacophoric features [Srinivasan et al. 2002].

In the Pharmacophore-Derived Queries (PDQ) method [Pickett et al. 1996], a much smaller 3-point pharmacophore key is generated based on six features (donor, acceptor, acid, base, aromatic centre and hydrophobe) and six distance bins (with each bin covering a distance range of 2.5–5 Å). This gives a total of 5,916 potential pharmacophores. The number of times each pharmacophore is matched is also recorded, so providing information beyond the simple binary information encoded in most bitstrings.

3.3   Other 3D Descriptors

A variety of other 3D descriptors have been developed, many of which have 2D counterparts. For example, 3D *topographical indices* can be calculated from the distance matrix of a molecule [Randić and Razinger 1995]; these are analogous to the topological indices which are generated from a 2D connection table. Similarly, *geometric atom pairs* have been developed as an extension of atom pair descriptors [Sheridan et al. 1996]. Other descriptors are based upon time-consuming calculations that are not generally applicable to large data sets. For example, it is possible to derive properties such as dipole moments and HUMO and LUMO energies from quantum mechanical calculations [Leach 2001]. 3D properties such as the electrostatic potential are used in QSAR methods, such as Comparative Molecular Field Analysis (CoMFA), described in Chapter 4. Other techniques also use such descriptions, but rely upon the relative alignment of molecules; these are considered in Chapter 5. Finally, several descriptors have found particular use in the prediction of ADMET properties; these will be discussed in Chapter 8.

4.   DATA VERIFICATION AND MANIPULATION

Having calculated a series of descriptors for a set of molecules it is often advisable to examine their characteristics prior to using them in an analysis. In particular it can be important to evaluate the distribution of values for a given descriptor and to check for correlations between different descriptors which could lead to over-representation of certain information. Manipulation of the data may also be required. This could involve a simple technique such as scaling to ensure that each descriptor contributes equally to the analysis or it may involve a more complex technique such as principal components analysis (PCA) that results in a new set of descriptors with more desirable characteristics.

4.1   Data Spread and Distribution

One of the most straightforward verification procedures that should be performed for any descriptor is to examine the spread of values for the data set. If the values show no variation then there is nothing to be gained from inclusion of the descriptor. In some cases it may be important that the values of a descriptor follow a particular distribution, often the normal distribution.

One quantity that can be used to assess the spread of a descriptor is the *coefficient of variation*. This is equal to the standard deviation (σ) divided by the mean ($\langle x \rangle$); these two quantities are given by the following equations:

$$\langle x \rangle = \frac{1}{N} \sum_{i=1}^{N} x_i \tag{3.20}$$

$$\sigma = \sqrt{\frac{1}{N} \sum_{i=1}^{N} (x_i - \langle x \rangle)^2} \equiv \sqrt{\frac{1}{N} \left[\sum_{i=1}^{N} (x_i^2) - \frac{1}{N} \left(\sum_{i=1}^{N} x_i \right)^2 \right]} \tag{3.21}$$

where there are N data points and x_i is the value of descriptor x for data point i. The larger the coefficient of variation the better the spread of values.

4.2   Scaling

Descriptors often have substantially different numerical ranges and it is important that they are *scaled* appropriately so that each descriptor has an equal chance of contributing to the overall analysis. Otherwise, a descriptor that has a large range of values will overwhelm any variation seen in a descriptor that has a small range of values and will bias the results. Scaling is also often referred to as *standardisation*. There are many ways to scale data. In *unit variance* scaling (also known as *auto-scaling*), each descriptor value is divided by the standard deviation for that descriptor across all observations (molecules). Each scaled descriptor then has variance of one.

Unit variance scaling is usually combined with *mean centring* in which the average value of a descriptor is subtracted from each individual value. In this way all descriptors are centred on zero and have a standard deviation of one:

$$x_i' = \frac{x_i - \langle x \rangle}{\sigma} \tag{3.22}$$

where x_i' is the new, transformed value of the original x_i. *Range scaling* uses a related expression in which the denominator equals the difference between the maximum and minimum values. This gives a set of new values that vary between -0.5 and $+0.5$.

4.3   Correlations

Correlations between the descriptors should also be checked as a matter of routine. If highly correlated descriptors are used in combination then the information that they characterise may be over-represented. Many correlations can be identified from simple scatterplots of pairs of descriptor values; ideally, the points will be distributed with no discernible pattern as illustrated schematically in Figure 3-5.

When many descriptors need to be considered then it is usually more convenient to compute a pairwise correlation matrix. This quantifies the degree of correlation between all pairs of descriptors. Each entry (i, j) in the correlation matrix is the correlation coefficient between the descriptors x_i and x_j. The correlation coefficient, r, is given by:

$$r = \frac{\sum_{k=1}^{N} \left[(x_{i,k} - \langle x_i \rangle)(x_{j,k} - \langle x_j \rangle) \right]}{\sqrt{\sum_{k=1}^{N} (x_{i,k} - \langle x_i \rangle)^2 \sum_{k=1}^{N} (x_{j,k} - \langle x_j \rangle)^2}} \tag{3.23}$$

The values of the correlation coefficient range from -1.0 to $+1.0$. A value of $+1.0$ indicates a perfect positive correlation; a plot of x_i versus x_j would

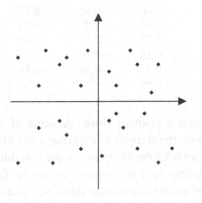

Figure 3-5. A pair of descriptors with no correlation will have points in all four quadrants of the scatter plot and with no obvious pattern or correlation.

Table 3-2. Descriptors of amino acids and amino acid side chains [El Tayar et al. 1992]. LCE and LCF are two lipophilicity constants of amino acid side chains, FET is the free energy of transfer of amino acid side chains from organic solvent into water, POL is a polarity parameter, VOL is the molecular volume of the amino acid and ASA is the solvent-accessible surface area of the amino acid.

Name	LCE	LCF	FET	POL	VOL	ASA
Ala	0.23	0.31	−0.55	−0.02	82.2	254.2
Arg	−0.79	−1.01	2.00	−2.56	163.0	363.4
Asn	−0.48	−0.60	0.51	−1.24	112.3	303.6
Asp	−0.61	−0.77	1.20	−1.08	103.7	287.9
Cys	0.45	1.54	−1.40	−0.11	99.1	282.9
Gln	−0.11	−0.22	0.29	−1.19	127.5	335.0
Glu	−0.51	−0.64	0.76	−1.43	120.5	311.6
Gly	0.00	0.00	0.00	0.03	65.0	224.9
His	0.15	0.13	−0.25	−1.06	140.6	337.2
Ile	1.2	1.80	−2.10	0.04	131.7	322.6
Leu	1.28	1.70	−2.00	0.12	131.5	324.0
Lys	−0.77	−0.99	0.78	−2.26	144.3	336.6
Met	0.90	1.23	−1.60	−0.33	132.3	336.3
Phe	1.56	1.79	−2.60	−0.05	155.8	366.1
Pro	0.38	0.49	−1.50	−0.31	106.7	288.5
Ser	0.00	−0.04	0.09	−0.40	88.5	266.7
Thr	0.17	0.26	−0.58	−0.53	105.3	283.9
Trp	1.85	2.25	−2.70	−0.31	185.9	401.8
Tyr	0.89	0.96	−1.70	−0.84	162.7	377.8
Val	0.71	1.22	−1.60	−0.13	115.6	295.1

Table 3-3. Correlation matrix for amino acid data.

Name	LCE	LCF	FET	POL	VOL	ASA
LCE	1.00	0.97	−0.93	0.74	0.37	0.46
LCF	0.97	1.00	−0.92	0.77	0.28	0.38
FET	−0.93	−0.92	1.00	−0.74	−0.26	−0.28
POL	0.74	0.77	−0.74	1.00	−0.35	−0.18
VOL	0.37	0.28	−0.26	−0.35	1.00	0.89
ASA	0.46	0.38	−0.28	−0.18	0.89	1.00

give a straight line with a positive slope. A value of −1.0 indicates a perfect negative correlation with the straight line having a negative slope. A value of zero indicates that there is no relationship between the variables. By way of illustration consider the set of amino acid descriptors shown in Table 3-2 [El Tayar et al. 1992]. The correlation matrix is given in Table 3-3; as can be seen there is a high degree of positive correlation between the two lipophilicity constants and between the volume and the solvent-accessible surface area. There is a strong negative correlation between the polarity parameter and the two lipophilicity parameters.

4.4 Reducing the Dimensionality of a Data Set: Principal Components Analysis

The *dimensionality* of a data set is the number of variables that are used to describe each object. *Principal Components Analysis* (PCA) is a commonly used method for reducing the dimensionality of a data set when there are significant correlations between some or all of the descriptors. PCA provides a new set of variables that have some special properties; it is often found that much of the variation in the data set can be explained by a small number of these principal components. The principal components are also much more convenient for graphical data display and analysis.

A simple 2D example will illustrate the concept of principal components. Consider the scatterplot shown in Figure 3-6. As can be seen, there is a high correlation between the x_1 and the x_2 values. Most of this variation can however be explained by introducing a single variable that is a linear combination of these (i.e. $z = x_1 - x_2$). The new variable (z) is referred to as a *principal component*.

In the general case of a multidimensional data set each of the principal components is a linear combination of the original variables or descriptors:

$$PC_1 = c_{1,1}x_1 + c_{1,2}x_2 + \cdots c_{1,p}x_p \tag{3.24}$$

$$PC_2 = c_{2,1}x_1 + c_{2,2}x_2 + \cdots c_{2,p}x_p \tag{3.25}$$

$$PC_i = c_{i,1}x_1 + c_{i,2}x_2 + \cdots c_{i,p}x_p = \sum_{j=1}^{p} c_{i,j}x_j \tag{3.26}$$

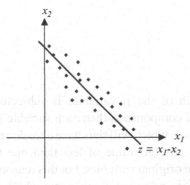

Figure 3-6. Simple illustration of principal components analysis.

where PC_i is the ith principal component and $c_{i,j}$ is the coefficient of the descriptor x_j, p being the number of descriptors. The first principal component maximises the variance in the data (i.e. the data have their greatest "spread" along this first principal component, as shown in Figure 3-6). The second principal component accounts for the maximum variance in the data that is not already explained by the first principal component, and so on for the third, fourth, etc. principal components. Another property of the principal components is that they are all orthogonal to each other. Thus in the simple 2D example shown in Figure 3-6 the second principal component would define an axis at right angles to the first.

The number of principal components that can be calculated is equal to the smaller of the number of data points (molecules) or the number of variables (descriptors) in the original data set. It is usually necessary to include all of the principal components to explain all of the variance in the data. However, in many data sets only a small number of principal components is required to explain a significant proportion of the variation, due to correlations between the original variables.

The principal components are calculated from the *variance–covariance matrix*; if A is the matrix with n rows (corresponding to n molecules) and p columns (for the p descriptors) then the variance–covariance matrix is the $n \times n$ matrix AA^T. The eigenvectors of this matrix are the coefficients $c_{i,j}$ of the principal components (a summary of matrices, eigenvalues and eigenvectors is provided in Appendix 1). The first principal component (the one that explains the largest amount of the variance) corresponds to the largest eigenvalue; the second principal component to the second largest eigenvalue and so on. The eigenvalues indicate the proportion of the variance that is explained by each of the principal components. If the eigenvalues are labelled λ_i then the first m principal components account for the following fraction of the total variation in the data set:

$$ f = \frac{\sum\limits_{i=1}^{m} \lambda_i}{\sum \lambda_i} \tag{3.27} $$

If, as is usual, each of the p variables is subjected to auto-scaling before extracting the principal components, then each variable will contribute a variance of one to the total data set, which will thus have a total variance of p. Any principal component which has an eigenvalue of less than one thus explains less of the variance than one of the original variables. For this reason it is common to consider as meaningful only those principal components which have eigenvalues greater than one. This is a widely applied rule-of-thumb when interpreting a PCA and when using the principal components in subsequent modelling, such as principal components regression (PCR) (to be discussed in Chapter 4).

By way of example, consider again the data in Table 3-2. As we observed some of these descriptors are highly correlated. A PCA of this data provides the following results:

$$PC_1 = 0.514\,LCE + 0.506\,LCF - 0.489\,FET + 0.371\,POL$$
$$+ 0.202\,VOL + 0.248\,ASA \quad \lambda_1 = 74.84 \quad v = 62\% \qquad (3.28)$$

$$PC_2 = -0.013\,LCE - 0.076\,LCF + 0.108\,FET - 0.477\,POL$$
$$+ 0.638\,VOL + 0.590\,ASA \quad \lambda_2 = 40.06 \quad v = 33\% \qquad (3.29)$$

$$PC_3 = 0.081\,LCE + 0.074\,LCF + 0.597\,FET + 0.373\,POL$$
$$- 0.365\,VOL + 0.600\,ASA \quad \lambda_3 = 3.47 \quad v = 2\% \qquad (3.30)$$

where v is the proportion of variance accounted for. Thus the first two principal components taken together account for 95% of the variance. The results of this PCA can be displayed graphically. Two of the most common ways are via *loadings plots* and *scores plots*. The loadings plot indicates the coefficients for each of the descriptors in the various principal components. The loadings plot for the first two principal components of the amino acid data set is shown in Figure 3-7. This shows the relative contribution of each descriptor to the different principal components. As can be seen the first principal component has reasonably significant contributions from all six descriptors whereas the second principal component is mostly composed of the volume, accessible surface area and polarity terms with little contribution from the lipophilicity and free energy of transfer terms. Moreover, the close proximity of the volume and accessible surface area

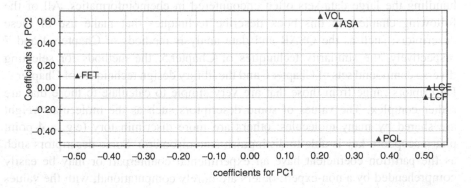

Figure 3-7. Loadings plot for the first two principal components of the amino acid data.

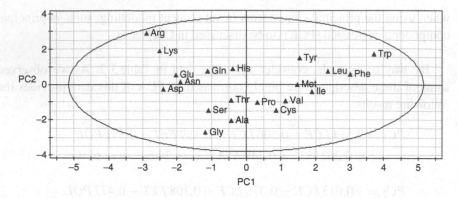

Figure 3-8. Scores plot showing the distribution of amino acids in the first two principal components.

terms and of the two lipophilicity constants reflects the high degree of correlation between these two pairs of descriptors.

The scores plot in Figure 3-8 shows how the various amino acids relate to each other in the space of the first two principal components. This plot indicates the underlying structure of the data; in this case it can be seen that similar amino acids are indeed grouped roughly together, with the charged and polar amino acids in one region (e.g. Arg, Lys, Glu, Asp), the small hydrophobic amino acids forming another group (e.g. Ala, Pro, Val) and the aromatic amino acids in a third region(e.g. Phe, Trp).

5.   SUMMARY

In this chapter, we have considered descriptors that are commonly used when handling the large data sets often encountered in chemoinformatics. All of the following chapters in this book describe techniques that make use of these descriptors, such as the QSAR and data analysis methods of Chapters 4 and 7 respectively, the similarity techniques of Chapter 5, the methods for selecting diverse compound sets of Chapter 6 and the library design techniques of Chapter 9. Descriptors range from those that are very simple to calculate to those that are rather complex. The values of some descriptors such as the molecular weight are shared by many molecules; others are more discriminatory (e.g. a 4-point pharmacophore key is able to distinguish stereoisomers). Some descriptors such as the partition coefficient have an experimental counterpart or may be easily comprehended by a non-expert; others are purely computational, with the values for new molecules being rather difficult to predict. As we shall see in subsequent chapters, selection of the appropriate set of descriptors is often critical to success.

Chapter 4

COMPUTATIONAL MODELS

1. INTRODUCTION

Most molecular discoveries today are the result of an iterative, three-phase cycle of design, synthesis and test. Analysis of the results from one iteration provides information and knowledge that enables the next cycle to be initiated and further improvements to be achieved. A common feature of this analysis stage is the construction of some form of model which enables the observed activity or properties to be related to the molecular structure. Many types of model are possible, with mathematical and statistical models being particularly common. Such models are often referred to as Quantitative Structure-Activity Relationships (QSARs) or Quantitative Structure–Property Relationships (QSPRs).

The focus in this chapter is on methods that can be applied to relatively small data sets. The analysis of large data sets, as might be obtained from high-throughput screening (HTS), is considered in Chapter 7 (note that many of the techniques discussed there can also be applied to small data sets). We will consider the use of simple and multiple linear regression for the derivation of QSAR models, including a discussion of the techniques that can be used to confirm the validity of such models. Also important is the design of a QSAR "experiment" including the construction of the data set and the descriptors employed. Finally, the techniques of PCR and partial least squares (PLS) are described.

2. HISTORICAL OVERVIEW

It is generally accepted that Hansch was the first to use QSARs to explain the biological activity of series of structurally related molecules [Hansch and Fujita 1964; Fujita et al. 1964; Hansch 1969]. Hansch pioneered the use of descriptors related to a molecule's electronic characteristics and to its hydrophobicity. This led him to propose that biological activity could be related to the molecular structure via equations of the following form:

$$\log(1/C) = k_1 \log P + k_2 \sigma + k_3 \tag{4.1}$$

75

where C is the concentration of compound required to produce a standard response in a given time, $\log P$ is the logarithm of the molecule's partition coefficient between 1-octanol and water and σ is the appropriate Hammett substitution parameter. This formalism expresses both sides of the equation in terms of free energy. An alternative formulation of this equation uses the parameter π, which as we discussed in Chapter 3 is the difference between the $\log P$ for the compound and the analogous hydrogen-substituted compound:

$$\log(1/C) = k_1\pi + k_2\sigma + k_3; \quad \pi = \log P_X - \log P_H \qquad (4.2)$$

Another important breakthrough was Hansch's proposal that the activity was parabolically dependent on $\log P$:

$$\log(1/C) = -k_1(\log P)^2 + k_2(\log P) + k_3\sigma + k_4 \qquad (4.3)$$

Hansch's rationale for suggesting the parabolic dependence on $\log P$ was that the drug's hydrophobicity should not be so low that the drug did not partition into the cell membrane nor so high that once in the membrane it simply remained there. An early example of such a non-linear relationship between a biological response and the partition coefficient is the following equation derived for the narcotic effect of thirteen barbiturates on mice [Hansch et al. 1968]:

$$\log(1/C) = -0.44(\log P)^2 + 1.58(\log P) + 1.93 \qquad (4.4)$$

The electronic characteristics of a molecule are also important in determining its activity against the target; the Hammett parameters provided a concise and convenient way to quantify these. Hammett and others had shown that for related compounds reaction rates (e.g. for the hydrolysis of benzoate esters) and positions of equilibrium (e.g. the ionisation constants of substituted benzoic acids) could be quantified using equations of the following form [Hammett 1970]:

$$\log(k/k_0) = \rho\sigma \qquad (4.5)$$

$$\log(K/K_0) = \rho\sigma \qquad (4.6)$$

These equations express the rate (k) or equilibrium (K) constant for a particular substituent relative to that for a reference compound (indicated using the subscript 0 and typically that for which the substituent is hydrogen). The substituent parameter σ is determined by the nature of the substituent and whether it is *meta* or *para* to the carboxylic acid or ester group on the aromatic ring. The reaction constant ρ is fixed for a particular process with the standard reaction being the dissociation of benzoic acids ($\rho = 1$). There have been many extensions to Hammett's original formulation, such as the work of Taft who showed how to incorporate steric effects. Another key development of particular relevance to the QSAR community was the suggestion by Swain and Lupton that σ values could be written as a weighted linear combination of two components: a field component (F) and a resonance component (R) [Swain and Lupton 1968; Swain et al. 1983]. Many phenomena had been found to follow a Hammett-style equation but were not well correlated using Hammett's original substitution values. The Swain–Lupton proposal enabled many of these substitution constant sets to be rationalised.

3. DERIVING A QSAR EQUATION: SIMPLE AND MULTIPLE LINEAR REGRESSION

Various mathematical methods have been used to derive QSAR models but perhaps the most widely used technique is linear regression. The simplest type of linear regression equation has the following form:

$$y = mx + c \tag{4.7}$$

In this equation, y is called the *dependent variable* with x being the *independent variable*. In QSAR or QSPR y would be the property that one was trying to model (such as the biological activity) and x would be a molecular descriptor such as $\log P$ or a substituent constant (in the rest of this chapter "variable" and "descriptor" are used interchangeably to refer to the independent variable). The aim of linear regression is to find values for the coefficient m and the constant c that minimise the sum of the differences between the values predicted by the equation and the actual observations. This is illustrated graphically in Figure 4-1. The values of m and c are given by the following equations (in which N is the number of data values):

$$m = \frac{\sum_{i=1}^{N} (x_i - \langle x \rangle)(y_i - \langle y \rangle)}{\sum_{i=1}^{N} (x_i - \langle x \rangle)^2} \tag{4.8}$$

$$c = \langle y \rangle - m \langle x \rangle \tag{4.9}$$

The line described by the regression equation passes through the point $(\langle x \rangle, \langle y \rangle)$ where $\langle x \rangle$ and $\langle y \rangle$ are the means of the independent and dependent

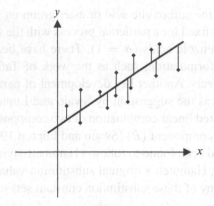

Figure 4-1. The aim of linear regression is to minimise the sum of the differences between the actual observations and those predicted from the regression equation (represented by the vertical lines from each point to the best-fit line).

variables respectively:

$$\langle x \rangle = \frac{1}{N} \sum_{i=1}^{N} x_i \tag{4.10}$$

$$\langle y \rangle = \frac{1}{N} \sum_{i=1}^{N} y_i \tag{4.11}$$

It is common for there to be more than one independent variable, in which case the method is referred to as *multiple linear regression* (the term *simple linear regression* applies where there is just one independent variable). The formulae for the coefficients in multiple linear regression are not so simple as Equations 4.8 and 4.9, but they are easily calculated using a standard statistical package.

3.1   The Squared Correlation Coefficient, R^2

The "quality" of a simple or multiple linear regression can be assessed in a number of ways. The most common of these is to calculate the squared correlation coefficient, or R^2 value (also written r^2). This has a value between zero and one and indicates the proportion of the variation in the dependent variable that is explained by the regression equation. Suppose $y_{calc,i}$ are the values obtained by feeding the relevant independent variables into the regression equation and y_i are the corresponding experimental observations. The following quantities can then be calculated:

$$\text{Total Sum of Squares, } TSS \quad = \sum_{i=1}^{N} (y_i - \langle y \rangle)^2 \tag{4.12}$$

$$\text{Explained Sum of Squares, } ESS = \sum_{i=1}^{N} (y_{\text{calc},i} - \langle y \rangle)^2 \tag{4.13}$$

$$\text{Residual Sum of Squares, } RSS \quad = \sum_{i=1}^{N} (y_i - y_{\text{calc},i})^2 \tag{4.14}$$

Thus,

$$TSS = ESS + RSS \tag{4.15}$$

R^2 is given by the following relationships:

$$R^2 = \frac{ESS}{TSS} \equiv \frac{TSS - RSS}{TSS} \equiv 1 - \frac{RSS}{TSS} \tag{4.16}$$

An R^2 of zero corresponds to a situation in which none of the variation in the observations is explained by the variation in the independent variables whereas a value of one corresponds to a perfect explanation. The R^2 statistic is very useful but taken in isolation it can be misleading. By way of illustration, Figure 4-2 shows five data sets for which the value of R^2 is approximately the same (0.7). With one exception (the data set shown top left) the single "best fit" straight line obtained from regression is clearly inappropriate, due to the presence of outliers or some other trend within the data.

3.2　　Cross-Validation

Cross-validation methods provide a way to try and overcome some of the problems inherent in the use of the R^2 value alone. Cross-validation involves the removal of some of the values from the data set, the derivation of a QSAR model using the remaining data, and then the application of this model to predict the values of the data that have been removed. The simplest form of cross-validation is the *leave-one-out* approach (LOO), where just a single data value is removed. Repeating this process for every value in the data set leads to a *cross-validated* R^2 (more commonly written Q^2 or q^2). The cross-validated R^2 value is normally lower than the simple R^2 but is a true guide to the predictive ability of the equation.

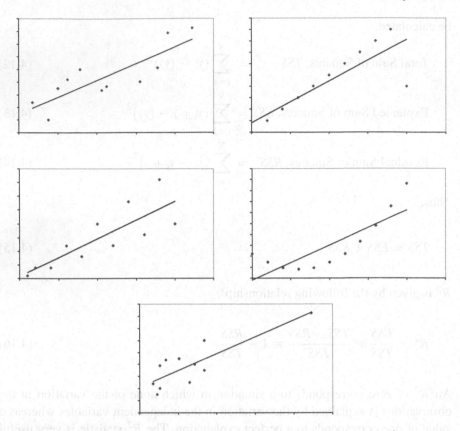

Figure 4-2. A variety of data sets which all have approximately the same R^2 value (0.7). (Adapted from [Livingstone 1995].)

Thus R^2 measures goodness-of-fit whereas Q^2 measures goodness-of-prediction. The simple LOO approach is widely considered to be inadequate and has largely been superseded by a more robust variant, in which the data set is divided into four or five groups, each of which is left out in turn to generate the Q^2 value. By repeating this procedure a large number of times (of the order of 100), selecting different groupings at random each time, a mean Q^2 can be derived. For a well-behaved system there should not be much variation in the equations obtained during such a procedure, nor should the R^2 value calculated using the entire data set be significantly larger than the mean Q^2. If there is a large discrepancy then it is likely that the data has been over-fit and the predictive ability of the equation will be suspect. A more rigorous procedure is to use an external set of molecules that is not used to build the model [Golbraikh and Tropsha 2002]. In some cases it may be possible to validate a model using temporal sets that contain molecules synthesised after the model is constructed. The *Predictive Residual Sum of Squares* (*PRESS*) is another measure of predictive ability. This is analogous to the *RSS* but rather than using values $y_{calc,i}$ calculated from the model *PRESS* uses predicted values $y_{pred,i}$; these are values for data not used to derive the model. Q^2 is related to *PRESS* as follows:

$$Q^2 = 1 - \frac{PRESS}{\sum\limits_{i=1}^{N} (y_i - \langle y \rangle)^2}; PRESS = \sum\limits_{i=1}^{N} (y_i - y_{pred,i})^2 \qquad (4.17)$$

The value of $\langle y \rangle$ in Equation 4.17 should strictly be calculated as the mean of the values for the appropriate cross-validation group rather than the mean for the entire data set, though the latter value is often used for convenience.

3.3 Other Measures of a Regression Equation

The standard error of prediction (s) is another commonly reported parameter; it indicates how well the regression function predicts the observed data and is given by:

$$s = \sqrt{\frac{RSS}{N - p - 1}} \qquad (4.18)$$

where p is the number of independent variables in the equation.

Two additional quantities used to assess a linear regression are the *F statistic* and the *t statistic*. *F* equals the *ESS* divided by the Residual Mean Square, as given by the following equation:

$$F = \frac{ESS}{p} \frac{N - p - 1}{RSS} \qquad (4.19)$$

The form of this equation reflects the number of *degrees of freedom* associated with each parameter. The *ESS* is associated with p degrees of freedom and the *RSS* with $N - p - 1$ degrees of freedom. Their respective means are therefore given by dividing by these quantities. A perhaps more easily comprehended illustration is the number of degrees of freedom associated with the *TSS*. This is $N - 1$ (rather than N) because the fitted line in a linear regression always passes through the mean of the dependent and independent variables ($\langle x \rangle, \langle y \rangle$) so reducing the number of degrees of freedom by one.

The calculated value for F is compared with values in statistical tables. These tables give values of F for different confidence levels. If the computed value is greater than the tabulated value then the equation is said to be significant at that particular confidence level. Higher values of F correspond to higher significance levels (i.e. greater confidence). For a given confidence level the threshold value of F falls as the number of independent variables decreases and/or the number of data points increases, consistent with the desire to describe as large a number of data points with as few independent variables as possible. The statistical tables provide values of F as a function of p and $(N - p - 1)$.

The t statistic indicates the significance of the individual terms in the linear regression equation. If k_i is the coefficient in the regression equation associated with a particular variable x_i then the t statistic is given by:

$$t = \left| \frac{k_i}{s\,(k_i)} \right| \; ; \; s\,(k_i) = \sqrt{\frac{RSS}{N-p-1} \cdot \frac{1}{\sum\limits_{j=1}^{N} \left(x_{i,j} - \langle x_i \rangle \right)^2}} \tag{4.20}$$

where $s(k_i)$ is known as the standard error of the coefficient. As with the F statistic, values of t computed from Equation 4.20 can be compared with values listed in statistical tables at various confidence levels; if the computed value is greater than the value in the table then the coefficient is considered significant.

4.   DESIGNING A QSAR "EXPERIMENT"

In order to derive a useful and predictive QSAR model it is necessary to design the "experiment" as carefully as one might design an experiment in the laboratory. The goal is generally to derive as parsimonious a model as possible; one that uses the smallest number of independent variables to explain as much data as possible. The following section provides some guidelines that may help avoid some of the common pitfalls.

The first guideline concerns the size of the data set. A general rule-of-thumb is that there should be at least five compounds for each descriptor that is included in the regression analysis. Second, simple verification checks should be performed on any descriptors proposed for inclusion in the analysis. These checks were described in Chapter 3; for example, if the values of a given descriptor show no variation then nothing is gained by its inclusion. The descriptor values should also in general have a good "spread", with no extreme outliers. It may be appropriate to scale the descriptors.

Correlations between the descriptors should also be checked, as described in Chapter 3. If pairs of correlated descriptors exist it is necessary to decide which to remove. One approach is to remove the descriptors that have the highest correlations with other descriptors. Conversely, one may choose to keep the descriptors that have the highest correlations in order to reduce the size of the data set as much as possible. Should two correlated descriptors each be correlated with the same number of other descriptors then the CORCHOP procedure removes that whose distribution deviates most from normal [Livingstone and Rahr 1989]. Alternatively one can avoid the problem altogether and use a data reduction technique such as PCA (see Chapter 3) to derive a new set of variables that are by definition uncorrelated.

4.1   Selecting the Descriptors to Include

Having identified a set of suitably distributed, non-correlated descriptors, it is necessary to decide which should be incorporated into the QSAR equation. Knowledge of the problem domain may suggest a subset of descriptors to consider. An alternative is to use an automated procedure. Two such procedures are *forward-stepping regression* and *backward-stepping regression*. Forward-stepping regression initially generates an equation containing just one variable (typically chosen to be the variable that contributes the most to the model, as might be assessed using the *t* statistic). Second, third and subsequent terms are added using the same criterion. Backward-stepping regression starts with an equation involving all of the descriptors which are then removed in turn (e.g. starting with the variable with the smallest *t* statistic). Both procedures aim to identify the equation that best fits the data, as might be assessed using the *F* statistic or with cross-validation. An extension of this approach is the use of GAs to generate a population of linear regression QSAR equations, each with different combinations of descriptors [Rogers and Hopfinger 1994]. The output from this procedure is a family of models from which one can select the highest scoring model or which can be analysed further to determine the most commonly occurring descriptors.

4.2   Experimental Design

Planning a QSAR "experiment" can extend even to a consideration of the synthesis of the compounds that will be used to build the model. Careful design may enable an effective QSAR to be determined from a minimal number of molecules and so may reduce the synthetic effort. For example, if it is believed that a particular descriptor is important in explaining the activity then it is clearly important to ensure that this property has a good spread of values in the set of molecules. Experimental design techniques can help to extract the maximum information from the smallest number of molecules. One such technique is *factorial design*, which is most conveniently illustrated with a simple example. Suppose there are two variables that might affect the outcome of an experiment. For example, the outcome (formally known as the *response*) may be the yield of a chemical synthesis and the variables (or *factors*) might be the reaction time, t, and the temperature, T. If each of these factors is restricted to two values, then there are four possible experiments that could be performed, corresponding to the four possible combinations (i.e. T_1t_1, T_1t_2, T_2t_1, T_2t_2). The second and third combinations provide information about changing one variable relative to the first combination whereas the fourth experiment involves a change in both variables; as such it might provide information on interactions between the variables. In general, if there are n variables, each restricted to two values, then a full factorial design involves 2^n experiments. A full factorial design can in principle provide information about all possible interactions between the factors, but this may require a large number of experiments. Fortunately, it is rare for interactions involving large numbers of factors to be significant; in general single factors are more important than pairwise interactions, which in turn are more important than combinations of three factors. A fractional factorial design involves fewer experiments and should still be able to identify the key factors, together with some information about interactions between factors.

Factors such as temperature and time can be independently controlled by the scientist. However, it may not be possible when formulating a QSAR to identify compounds with all of the desired combinations of descriptor values that a factorial analysis might suggest. This may be due to inherent correlations between certain types of molecular properties. For example, molecular weight and the molecular surface area are approximately correlated making it difficult to find molecules with a low molecular weight and a high surface area and *vice versa*. Under such situations other techniques are available for the construction of "well-balanced" sets. *D-optimal design* is one such technique. This method uses the variance–covariance matrix; if $\mathbf{A}$ is the matrix with n rows (corresponding to n molecules) and p columns (for the p descriptors) then the variance–covariance matrix is the $n \times n$ matrix $\mathbf{AA}^T$. D-optimal design aims to identify a subset of n molecules from a larger set of compounds such that the determinant of this variance–covariance matrix for the subset is a maximum. A Monte Carlo search approach can be used to search for D-optimal subsets which should have maximum variance (i.e. a large spread of values for the descriptors) and minimum covariance (i.e. small degree of correlation between the descriptors).

Multiple linear regression is a *supervised* learning method because it makes use of the dependent data (i.e. *y*) in order to derive a model. PCA is an example of an *unsupervised* learning method because it does not use information about the dependent variable when constructing the model. An additional distinction may be drawn between *parametric* techniques and *non-parametric* techniques. A parametric technique relies on the variables being distributed according to some specific distribution (usually the Gaussian distribution). As such it is possible to rely upon the properties of that particular distribution when deriving the statistical model (as is the case for linear regression). A non-parametric technique by contrast does not rely on the properties of some predetermined distribution.

4.3   Indicator Variables

Many descriptors have been used in QSAR equations. The first QSAR equations were often derived for structurally related series of compounds that differed at just one substitution point. This enabled parameters such as the Hammett substitution constants to be used successfully. Many substituent-based descriptors were introduced to represent other properties relevant to biological activity [Silipo and Vittoria 1990]. However, equations based on such descriptors are only valid within that chemical series. There is an increasing trend towards the simultaneous variation at multiple substitution points (as might, for example, be produced by combinatorial chemistry) and the progression of structurally dissimilar chemical series. In order to generate QSAR equations that are applicable to different chemical series it may be necessary to use whole-molecule descriptors such as those described in Chapter 3. An alternative is to use *indicator variables* to distinguish different members of a data set. Indicator variables are most often used to indicate the presence or absence of particular chemical features and usually take one of two values (zero or one). For example, they may be used to produce an equation that is applicable to both *meta-* and *ortho*-substituted systems, or for a set of compounds where some contain a particular functional group and some do not. An example is the following equation for the binding of sulphonamides of the type $X–C_6H_4–SO_2NH_2$ to human carbonic anhydrase [Hansch et al. 1985]:

$$\log K = 1.55\sigma + 0.64\log P - 2.07I_1 - 3.28I_2 + 6.94 \tag{4.21}$$

where I_1 takes the value one for *meta* substituents (zero for others) and I_2 is one for *ortho* substituents (zero for others).

The regression coefficients for indicator variables are calculated in just the same way as for the other descriptors. An advantage of using indicator variables is that they may permit data sets for molecules that act by the same underlying mechanism to be combined into a larger set which may in turn give rise to more reliable and predictive models. This is illustrated above where a single equation can account for both *meta* and *ortho* substituted inhibitors.

4.4   Free–Wilson Analysis

Free–Wilson analysis is a technique devised in the early days of QSAR [Free and Wilson 1964] that is related to the use of indicator variables. Starting from a series of substituted compounds with their associated activity, the aim as usual is to generate an equation of the following form:

$$y = a_1 x_1 + a_2 x_2 + a_3 x_3 + \cdots \qquad (4.22)$$

where x_1, x_2, etc. correspond to the various substituents at the different positions in the series of structures. These x_i variables are essentially equivalent to indicator variables and they take a value of zero or one to indicate the presence or absence of a particular substituent at the relevant position in the compound. Standard multiple linear regression methods are used to derive the coefficients a_i, which indicate the contribution of the corresponding substituent/position combination to the activity.

A key advantage of the Free–Wilson method is its simplicity. Moreover, it does not require any substituent parameters or other descriptors to be defined; only the activity is needed. However, it is based on certain assumptions that may not be valid for some data sets. First, it is assumed that each substituent on the parent structure makes a constant contribution to the activity. Second, it is assumed that these contributions are additive, and third that there are no interactions between the substituents nor between the substituents and the parent structure. In addition to these limitations there may also be problems when the same sets of substituents always occur together in the molecules. Another drawback is that it is not possible to make predictions for molecules containing substituents not included in the data set.

4.5   Non-Linear Terms in QSAR Equations

One of the key advances introduced by Hansch was the use of a QSAR equation dependent on the square of log P. A qualitative explanation for this parabolic dependence was provided in the introduction to this chapter; Hansch also showed how such a relationship could be derived in a more quantitative way [Hansch 1969.

A number of alternative non-linear models have also been described. One of these alternatives is the *bilinear model* [Kubinyi 1976, 1977] in which the ascending and descending parts of the function have different slopes (unlike the parabolic model, which is symmetrical). This may more closely mimic the observed data. The bilinear model uses an equation of the following form:

$$\log(1/C) = k_1 \log P - k_2 (\log(\beta P + 1)) + k_3 \tag{4.23}$$

In general, the Hansch parabolic model is often more appropriate to model complex assays where the drug must cross several barriers whereas the bilinear model may be more suited to simpler assay systems.

4.6 Interpretation and Application of a QSAR Equation

The ultimate value of a QSAR model, of course, depends on the impact that it has on the direction of the project. The most obvious use of a QSAR is in predicting the activity of molecules not yet synthesised. Prediction can be interpolative or extrapolative [Tute 1990]. An interpolative prediction is one within the range of properties of the set of molecules used to derive the QSAR; an extrapolative prediction is one beyond it. Interpolative prediction is generally more reliable than extrapolative prediction. Inevitably, perhaps, one is most interested in predictions that improve the potency or other biological properties of the molecules and these are often extrapolative (e.g. if the correlation is linear with the physicochemical parameters). Nevertheless, there are many examples where this has been possible [Fujita 1990]. In general, one should proceed gradually until the maximum potency can be achieved. In addition to moving a project forward, a QSAR model may also help one to determine when a particular series is unlikely to lead to the desired goal. Resources can therefore be shifted to explore other areas.

There have been some efforts to quantify the accuracy of the predictions from QSAR models. This question often arises in the context of "global" and "local" models [Brown and Lewis 2006]. The data sets used for global models are often large and heterogeneous whereas those for local models tend to be smaller and more homogeneous. Many of the earlier applications of QSAR techniques were best considered as local models, being restricted to single chemical series, whereas there is now significant interest in generating global models, often of properties relating to physicochemical and ADME phenomena. The concept of *model domain* has proved to be useful when assessing the prediction accuracy of such global models. A simple but effective measure was found to be the similarity to the molecules in the training set, as measured by the distance to the nearest neighbour and the number of nearest neighbours [Sheridan et al. 2004]. An extension of these ideas enables one to apply a quantitative correction factor based on this "distance to model" to the prediction from a QSAR model and thereby improve its accuracy [Xu and Gao 2003; Todeschini et al. 2004; Bruneau and McElroy 2006].

A QSAR model may provide insights into the factors relevant for the activity of a series of molecules and the mechanisms underlying the activity. Indeed, this is often the most important part of a QSAR analysis. As one might expect, most benefit derives from models based on chemically meaningful and comprehensible descriptors. The non-linear dependency on $\log P$ as originally introduced by Hansch is one such example. Another example is a QSAR study on the inhibition of alcohol dehydrogenase (ADH) by 4-substituted pyrazoles (Figure 4-3) [Hansch et al. 1986; Hansch and Klein 1986]. In this case it was possible to confirm the interpretation of the QSAR equation by subsequent examination of the protein–ligand x-ray structures. The following QSAR equation was derived for the inhibition of rat liver ADH:

$$\log 1/K_i = 1.22 \log P - 1.80\sigma_{meta} + 4.87 \tag{4.24}$$

where σ_{meta} is the Hammett constant for *meta* substitutents and K_i is the enzyme inhibition constant. The negative Hammett parameter is consistent with the fact that the pyrazole group binds to the catalytic zinc atom in the enzyme; electron-releasing substituents X will increase the electron density on the nitrogen and increase binding. The near-unity coefficient of $\log P$ was interpreted as meaning that the partitioning of X between water and the enzyme parallels that between water and octanol, with complete desolvation of the substituent. The x-ray structure revealed that the substituent occupied a long, channel-like hydrophobic pocket and that there would therefore indeed be expected to be complete desolvation on binding.

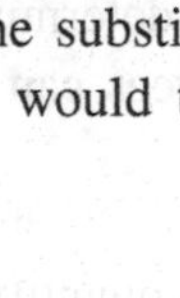

Figure 4-3. Generic structure of 4-pyrazole inhibitors of alcohol dehydrogenase.

5.   PRINCIPAL COMPONENTS REGRESSION

In *principal components regression* (PCR) the principal components are themselves used as variables in a multiple linear regression. As most data sets provide many fewer "significant" principal components than variables (e.g. principal components whose eigenvalues are greater than one) this may often lead to a concise QSAR equation of the form:

$$y = a_1 PC_1 + a_2 PC_2 + a_3 PC_3 + \cdots \qquad (4.25)$$

This equation is now expressed in terms of the principal components, rather than the original variables. A key point about PCR is that the most important principal components (those with the largest eigenvalues) are not necessarily the most important to use in the PCR equation. For example, if forward-stepping variable selection is used then the principal components will not necessarily be chosen in the order one, two, three, etc. This is because the principal components are selected according to their ability to explain the variance in the independent variables whereas PCR is concerned with explaining the variation in the dependent (y) variable. Nevertheless, it is frequently the case that at least the first two principal components will give the best correlation with the dependent variable. Another feature of PCR is that as extra terms are added the regression coefficients do not change. This is a consequence of the orthogonality of the principal components; the extra principal components explain variance in the dependent variable not already explained by the existing terms. One drawback of PCR is that it may be more difficult to interpret the resulting equations, for example to decide which of the original molecular properties should be changed in order to enhance the activity.

6.   PARTIAL LEAST SQUARES

The technique of *partial least squares* [Wold 1982] is similar to PCR, with the crucial difference that the quantities calculated are chosen to explain not only the variation in the independent (x) variables but also the variation in the dependent (y) variables as well. Partial least squares is usually abbreviated to PLS; an alternative name for the technique is *projection to latent structures*. PLS expresses the dependent variable in terms of quantities called latent variables, comparable to the principal components in PCR. Thus:

$$y = a_1 t_1 + a_2 t_2 + a_3 t_3 + \cdots a_n t_n \qquad (4.26)$$

The latent variables t_i are themselves linear combinations of the independent variables (x_i):

$$t_1 = b_{11}x_1 + b_{12}x_2 + \cdots b_{1p}x_p \qquad (4.27)$$
$$t_2 = b_{21}x_1 + b_{22}x_2 + \cdots b_{2p}x_p \qquad (4.28)$$
$$t_i = b_{i1}x_1 + b_{i2}x_2 + \cdots b_{ip}x_p \qquad (4.29)$$

The number of latent variables that can be generated is the smaller of the number of variables or the number of observations. As with PCA the latent variables are orthogonal to each other. However, when calculating the latent variables PLS takes into account not only the variance in the x variables but also how this corresponds to the values of the dependent variable.

The first latent variable t_1 is a linear combination of the x values that gives a good explanation of the variance in the x-space, similar to PCA. However, it is also defined so that when multiplied by its corresponding coefficient a_1 it provides a good approximation to the variation in the y values. Thus a graph of the observed values y against a_1t_1 should give a reasonably straight line. The differences between these observed and predicted values are often referred to as the *residuals*; these represent the variation not explained by the model.

The second latent variable is then determined. This second latent variable is orthogonal to the first and enables more of the variation in the x dimension to be described. It also enables the variation in the y values to be further improved, with smaller residuals; a graph of y versus $a_1t_1 + a_2t_2$ will show an even better correlation than was the case for just the first latent variable.

To illustrate the use of PLS we again consider the amino acid data set introduced in Chapter 3 in our discussion of correlations and PCA. Free energies of unfolding have been reported for a series of mutants of the tryptophan synthase α subunit in which a single amino acid substitution had been made [Yutani et al. 1987]. These data (for 19 amino acids) are reproduced in Table 4-1.

PLS analysis on these 19 data points using the descriptors in Table 3-2 (Chapter 3) suggests that just one component is significant ($R^2 = 0.43$, $Q^2 = 0.31$); although the R^2 can be improved to 0.52 by the addition of a second component the Q^2 value falls to 0.24. By comparison, the equivalent multiple linear regression model with all six variables has an R^2 of 0.70 but the Q^2 is low (0.142). The quality of the one-component PLS model can be assessed graphically by plotting the predicted free energies against the measured values as shown in Figure 4-4.

Table 4-1. Free energies of unfolding of tryptophan synthase α subunit mutants.

Amino acid	Free energy of unfolding (kcal/mol)
Ala	8.5
Asn	8.2
Asp	8.5
Cys	11.0
Gln	6.3
Glu	8.8
Gly	7.1
His	10.1
Ile	16.8
Leu	15.0
Lys	7.9
Met	13.3
Phe	11.2
Pro	8.2
Ser	7.4
Thr	8.8
Trp	9.9
Tyr	8.8
Val	12.0

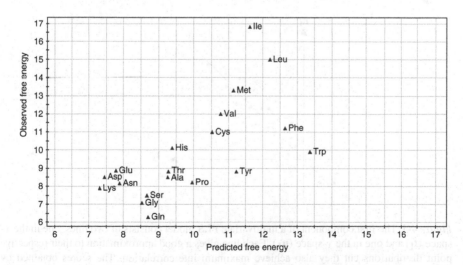

Figure 4-4. Graph of observed versus predicted free energies of unfolding for a one-component PLS model based on the amino acid descriptors and data in Tables 3-2 and 4-1.

As can be seen from the graph in Figure 4-4 this initial PLS model is rather poor; in particular the aromatic amino acids have a much worse fit than the others. Removal of these data and incorporation of some squared terms provides a much improved model as is discussed elsewhere [Wold 1994].

One important feature of PLS is its ability to cope with data sets with more than one dependent variable (i.e. multivariate problems). Here, the first component can be considered to constitute two vectors, one in the space of the x-variables (i.e. t_1) and one in the y-space (usually written u_1). These two lines give good approximations to their respective data points but also have maximum correlation with each other (Figure 4-5).

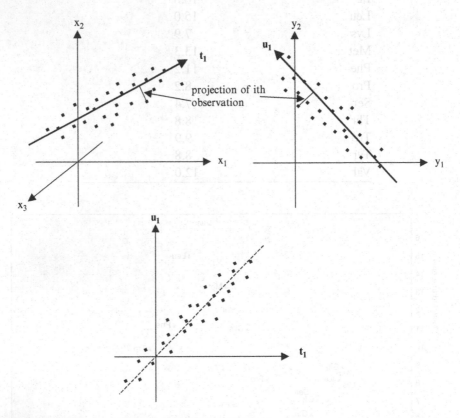

Figure 4-5. The first component of a multivariate PLS model consists of two lines, one in the x-space (t_1) and one in the y-space (u_1). Each line gives a good approximation to their respective point distributions but they also achieve maximum intercorrelation. The scores obtained by projecting each point onto these lines can be plotted (bottom figure); if there a perfect match between the x and y data all the points would lie on the diagonal of slope one.

An early illustrative example of a multivariate analysis using PLS is the study by Dunn et al. (1984) who examined a set of eight halogenated hydrocarbons for which five different measures of toxicity had been determined. Each molecule was characterised by 11 descriptors, comprising both whole-molecule properties such as $\log P$, molecular weight and molar refractivity together with calculated parameters related to the electronic structure. A PLS analysis on seven of the compounds provided three significant components; this model was then used to predict the activities for the eighth compound.

An important consideration when performing PLS analyses, and a matter of considerable (and continuing) debate concerns the number of latent variables to include in a PLS model. The fit of the model (i.e. R^2) can always be enhanced by the addition of more latent variables. However, the predictive ability (i.e. Q^2) will typically only show such an improvement in the initial stages; eventually Q^2 either passes through a maximum or else it reaches a plateau. This maximum in Q^2 corresponds to the most appropriate number of latent variables to include. Some practitioners use the sPRESS parameter to guide the selection of latent variables; this is the standard deviation of the error of the predictions:

$$s_{PRESS} = \sqrt{\frac{PRESS}{N - c - 1}} \qquad (4.30)$$

where c is the number of PLS components in the model and N is the number of compounds. It has been suggested that one should use the smallest number of latent variables that gives a reasonably high Q^2 and that each of these latent variables should produce a fall in the value of sPRESS of at least 5% [Wold et al. 1993]. Another related parameter is $SDEP$, the standard error of prediction:

$$SDEP = \sqrt{\frac{PRESS}{N}} \qquad (4.31)$$

The different denominators in sPRESS and $SDEP$ mean that the latter does not penalise an increased number of components (c) and so a model selected using $SDEP$ may have more components than a model selected using sPRESS.

Other techniques can be used to demonstrate the predictivity and stability of a PLS model (and are also applicable to other QSAR methods). *Bootstrapping* attempts to simulate the effect of using a larger data set by randomly selecting a number of different sets, each of size N, from the data set. In each randomly selected set some of the data will be present more than once and some of the data will not be present at all. By deriving PLS models for each of these randomly selected sets one can assess the variation in each of the terms in the PLS model. Another approach is to randomly assign the activities, such that the wrong y value is associated with each row of x values. A PLS model is then derived for this set and the procedure repeated. The predictive nature of the model associated with the true, non-randomised, data set should be significantly greater than is obtained for the random sets.

7.   MOLECULAR FIELD ANALYSIS AND PARTIAL LEAST SQUARES

One of the most significant developments in QSAR in recent years was the introduction of *Comparative Molecular Field Analysis* (CoMFA) by Cramer et al. [1988] and Kubinyi [1998a]. The aim of CoMFA is to derive a correlation between the biological activity of a series of molecules and their 3D shape, electrostatic and hydrogen-bonding characteristics. The data structure used in a CoMFA analysis is derived from a series of superimposed conformations, one for each molecule in the data set. These conformations are presumed to be the biologically active structures, overlaid in their common binding mode. Each conformation is taken in turn, and the molecular fields around it are calculated. This is achieved by placing the structure within a regular lattice and calculating the interaction energy between the molecule and a series of probe groups placed at each lattice point. The standard fields used in CoMFA are electrostatic and steric. These are calculated using a probe comprising an sp^3-hybridised carbon atom with a charge of $+1$. The electrostatic potential is calculated using Coulomb's law and the steric field using the Lennard–Jones potential (see Appendix 2). This process is illustrated in Figure 4-6; the energy calculations are analogous to the GRID approach devised by Goodford for use in structure-based design [Goodford 1985]. The results of these calculations are placed into a data matrix. Each row in the data matrix corresponds to one molecule and each column corresponds to the interaction energy of a particular probe at a specific lattice point. If there are L lattice points and P probe groups, then there are $L \times P$ columns. A final column is added corresponding to the activities of the molecules (Figure 4-6).

PLS is then used to analyse the data matrix, the aim being to derive an equation that correlates the biological activity with the different probes at the various lattice points. The general form of the equation that results can be written:

$$\text{activity} = C + \sum_{i=1}^{L} \sum_{j=1}^{P} c_{ij} S_{ij} \qquad (4.32)$$

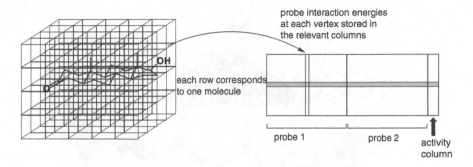

Figure 4-6. In CoMFA each molecule is placed within a regular grid and interaction energies are computed with probe groups to give a rectangular data structure.

Each coefficient c_{ij} corresponds to placing probe j at lattice point i with S_{ij} being the corresponding energy value. C is a constant. These coefficients indicate the significance of placing the relevant probe group at that lattice point. By connecting points with the same coefficients a 3D contour plot can be produced that identifies regions of particular interest. For example, one may be able to identify regions where it would be favourable to place positively charged groups or where increases in steric bulk would decrease the activity. This is illustrated in Figure 4-7 which shows the results of a CoMFA study on a series of coumarin substrates and inhibitors of cytochrome $P_{450}2A5$, an enzyme involved in the metabolism of drug substances [Poso et al. 1995]. Several different models were considered, with the highest Q^2 value being obtained for a five-component model based on the standard steric and electrostatic fields, though a model with a lower s_{PRESS} value and just three components was obtained using a field based on the lowest unoccupied molecular orbital (LUMO). Graphical display of the fields made it possible to identify regions around the ligands where it would be favourable and/or unfavourable to place negative charge and steric bulk. In particular, a large electrostatic field was observed near the lactone moiety, possibly indicating a hydrogen bonding interaction between the ligand and the enzyme (Figure 4-7).

The results of a 3D QSAR depend on a number of factors, each of which must be carefully considered. As with the derivation of other QSAR models one first needs to decide which compounds to include. It is then necessary to decide upon the probes to use, the energy function used to calculate the probe energies and the lattice spacing. One of the most important considerations involves the selection of the conformations and their alignment prior to the analysis. This may be relatively straightforward when one is working with a congeric series of compounds that all have some key structural feature that can be overlaid. For example, the original CoMFA paper examined a series of steroid molecules which were overlaid using the steroid nucleus. Data sets that contain molecules from different chemical series

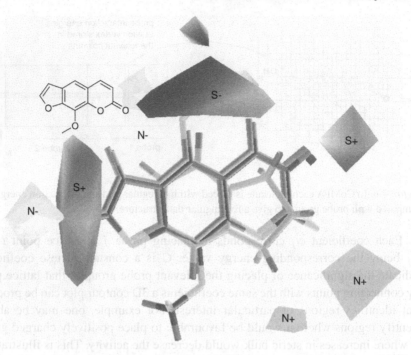

Figure 4-7. Contour representation of the key features from a CoMFA analysis of a series of coumarin substrates and inhibitors of cytochrome $P_{450}2A5$ [Poso et al. 1995]. N+ indicates regions where it is favourable to place a negative charge and N− where it would be unfavourable. Similarly, S+/S− indicate regions where it would be favourable/unfavourable to place steric bulk. Also shown are a number of superimposed ligands, and the chemical structure of a typical ligand, methoxsalen (top left). (Adapted from Leach 2001.)

may be more difficult to overlay. Another issue concerns variable selection. This is particularly acute for CoMFA due to the high degree of correlation between many of the variables, particularly those lattice points that are located in the same region of space [Cruciani et al. 1993]. A procedure called Generating Optimal Linear PLS Estimations (GOLPE) can be used to select multiple combinations of variables using fractional factorial design [Baroni et al. 1993]. For each combination of variables a PLS model is derived. At the end of the process only those variables that are found to significantly improve the predictive ability of the model are retained for the final PLS analysis. The basis of a related technique termed Smart Region Definition (SRD) [Pastor et al. 1997] involves first identifying the most important lattice points in a traditional CoMFA (these are the points with a high weight in the CoMFA equation). These points may themselves be merged together if they are spatially close and contain the same information (i.e. are correlated) to give a final subset of data points for the PLS analysis. By focussing on a relatively small number of information-rich, uncorrelated variables the final PLS model should be of higher quality than would otherwise be the case.

Since its introduction the CoMFA method has been extensively applied, leading to many hundreds of publications, reviews and books. In addition to the many practical applications of CoMFA much effort has also been expended investigating the theoretical foundations of the approach. These latter studies have identified a number of potential sources of error in the method and have resulted in recommendations for minimising such errors [Thibaut et al. 1993]. Several of these have already been mentioned, such as the selection and alignment of the active conformations, the grid spacing, grid orientation and selection of probes, the PLS procedure, the design of training and test sets and the cross-validation of the models. In addition, many variants and related techniques have been described; of these we will mention just one, Comparative Molecular Similarity Indices Analysis (CoMSIA) [Klebe et al. 1994]. The energy functions typically used to calculate the field values in CoMFA can give rise to significant variation in the energy for very small changes in position. For example, the Lennard–Jones potential traditionally used to calculate the steric field becomes very steep close to the van der Waals surface of the molecule and shows a singularity at the atomic positions (as does the Coulomb potential used to calculate the electrostatic field). In CoMFA, these issues are dealt with by applying arbitrary cut-offs; the overall effect is a loss of information, with the resulting contour plots being fragmented and difficult to interpret. In CoMSIA such fields are replaced by similarity values at the various grid positions. The similarities are calculated using much smoother potentials that are not so steep as the Lennard–Jones and Coulomb functions and have a finite value even at the atomic positions. The use of these similarity indices is considered to lead to superior contour maps that are easier to interpret. They also have the advantage of extending to regions within the ligand skeletons.

8.   SUMMARY

Mathematical models of biological processes continue to play an important role in drug discovery. In this chapter we have described multiple linear regression and PLS, which are currently the most widely used techniques for deriving QSAR and QSPR models. For example, many of the models for predicting ADMET properties use these techniques, as described in Chapter 8. Several other methods for constructing models are also available, some of which have particular utility for certain types of problems. In particular, methods that are more applicable to large data sets will be considered in Chapter 7 in the context of the analysis of HTS data.

Chapter 5

SIMILARITY METHODS

1. INTRODUCTION

Substructure and 3D pharmacophore searching involve the specification of a precise query, which is then used to search a database to identify compounds of interest. These approaches are clearly very useful ways to select compounds but they do have some limitations. First, one requires sufficient knowledge to be able to construct a meaningful substructure or 3D pharmacophore. However, this knowledge may not always be available if, for example, only one or two weakly active compounds are known. Second, in such searches a molecule either matches the query or it does not; as a consequence, the database is partitioned into two different sets with no sense of the relative ranking of the compounds. Finally, the user has no control over the size of the output. Thus, a query that is too general can result in an enormous number of hits whereas a very specific query may retrieve only a very small number of hits.

Similarity searching offers a complementary alternative to substructure searching and 3D pharmacophore searching. Here, a query compound is used to search a database to find those compounds that are most similar to it. This involves comparing the query with every compound in the database in turn. The database is then sorted in order of decreasing similarity to the query.

Similarity searching offers several advantages. First, there is no need to define a precise substructure or pharmacophore query since a single active compound is sufficient to initiate a search. Second, the user has control over the size of the output as every compound in the database is given a numerical score that can be used to generate a complete ranking. Alternatively, one can specify a particular level of similarity and retrieve just those compounds that exceed the threshold. Finally, similarity searching facilitates an iterative approach to searching chemical databases since the top-scoring compounds resulting from one search can be used as queries in subsequent similarity searches.

The rationale for similarity searching lies in the *similar property principle* [Johnson and Maggiora 1990] which states that structurally similar molecules tend to have similar properties. Thus, given a molecule of known biological activity, compounds that are structurally similar to it are likely to exhibit the same activity. This characteristic has been referred to as *neighbourhood behaviour* [Patterson et al. 1996]. In a typical similarity search, a molecule which is known to possess some desirable activity is used as a query in the hope of identifying other molecules that show the same activity. Similarity searching is thus widely used to identify compounds for screening. In fact, similarity searching is sometimes included within the family of techniques now known as *virtual screening*, which is the computational, or *in silico*, equivalent of experimental screening. Other virtual screening techniques are discussed in Chapter 8.

There are many examples of series of active compounds that provide evidence for the neighbourhood principle. A simple example is given in Figure 5-1 which shows three compounds which are all active at opioid receptors: morphine, codeine and heroin. The structural similarities of the compounds should be clear.

The main difficulty with similarity searching is that assessing the degree of similarity between two objects is subjective, with no "hard and fast" rules. In order to be able to quantify the similarity between two molecules, a similarity searching method requires two components: first, a set of numerical descriptors that can be used to compare molecules, and second, a similarity coefficient which provides a way of quantifying the degree of similarity based on the descriptors.

Similarity searching in chemical databases was first introduced in the mid-1980s [Carhart et al. 1985; Willett et al. 1986]. We begin our discussion by describing one of these early approaches which was based on 2D fingerprints and the use of the Tanimoto similarity coefficient. Many different methods have been developed subsequently involving a wide range of descriptors and a variety of similarity coefficients. The main body of the chapter is concerned with the various methods that have been developed over the years, starting with other types of 2D descriptors and then moving on to 3D similarity methods. Given the wide variety of different methods that has been developed it is important that their relative

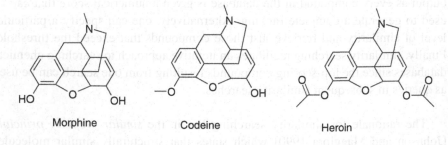

Morphine   Codeine   Heroin

Figure 5-1. Compounds active at opioid receptors.

performances can be compared. The final section describes some comparative studies that have been reported in the literature.

2. SIMILARITY BASED ON 2D FINGERPRINTS

Perhaps the most commonly used similarity method is based on the 2D fingerprints that were originally developed for substructure searching and which have already been described in Chapters 1 and 3. To recapitulate, fingerprints are binary vectors where each bit indicates the presence ("1") or absence ("0") of a particular substructural fragment within a molecule. The similarity between two molecules represented by 2D binary fingerprints is most frequently quantified using the Tanimoto coefficient which gives a measure of the number of fragments in common between the two molecules [Willett et al. 1986].

The Tanimoto similarity between molecules A and B, represented by binary vectors, is given by:

$$S_{AB} = \frac{c}{a + b - c} \tag{5.1}$$

where there are a bits set to "1" in molecule A, b bits set to "1" in molecule B, and c "1" bits common to both A and B. The value of the Tanimoto coefficient ranges from zero to one. A value of one indicates that the molecules have identical fingerprint representations (note that this does not necessarily mean they are identical molecules) and a value of zero indicates that there is no similarity (i.e. there are no bits in common between the two molecules). Figure 5-2 provides a simple hypothetical example of the calculation of the Tanimoto similarity coefficient.

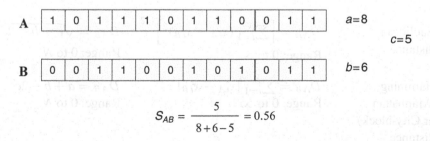

Figure 5-2. Calculating similarity using binary vector representations and the Tanimoto coefficient.

3. SIMILARITY COEFFICIENTS

The concept of similarity is important in many fields (not just chemistry) and many different similarity coefficients have been developed [Sneath and Sokal 1973; Willett et al. 1998]. Those most commonly used in chemical applications are shown in Table 5-1. As mentioned, the Tanimoto coefficient is the most widely used similarity coefficient for binary fingerprints such as structural keys and hashed fingerprints; it can also be used to calculate the similarity between molecules that are represented by continuous data such as the topological indices introduced in Chapter 3. Table 5-1 includes the more general form of the Tanimoto coefficient when the molecules A and B are represented by vectors, $\mathbf{x}$, of length N with the ith property having the value x_i. It is trivial to show that the continuous

Table 5-1. Similarity (S) or distance (D) coefficients in common use for similarity searching in chemical databases. For binary data a is defined as the number of bits set to "1" in molecule A, b as the number of bits set to "1" in molecule B and c as the number of bits that are "1" in both A and B. (From Willett et al. 1998.)

Name	Formula for continuous variables	Formula for binary (dichotomous) variables
Tanimoto (Jaccard) coefficient)	$S_{AB} = \dfrac{\sum_{i=1}^{N} x_{iA}x_{iB}}{\sum_{i=1}^{N} (x_{iA})^2 + \sum_{i=1}^{N} (x_{iB})^2 - \sum_{i=1}^{N} x_{iA}x_{iB}}$ Range: -0.333 to $+1$	$S_{AB} = \dfrac{c}{a+b-c}$ Range: 0 to 1
Dice coefficient (Hodgkin index)	$S_{AB} = \dfrac{2\sum_{i=1}^{N} x_{iA}x_{iB}}{\sum_{i=1}^{N} (x_{iA})^2 + \sum_{i=1}^{N} (x_{iB})^2}$ Range: -1 to $+1$	$S_{AB} = \dfrac{2c}{a+b}$ Range: 0 to 1
Cosine similarity (Carbó index)	$S_{AB} = \dfrac{\sum_{i=1}^{N} x_{iA}x_{iB}}{\left[\sum_{i=1}^{N} (x_{iA})^2 \sum_{i=1}^{N} (x_{iB})^2\right]^{1/2}}$ Range: -1 to $+1$	$S_{AB} = \dfrac{c}{\sqrt{ab}}$ Range: 0 to 1
Euclidean distance	$D_{AB} = \left[\sum_{i=1}^{N} (x_{iA} - x_{iB})^2\right]^{1/2}$ Range: 0 to ∞	$D_{AB} = \sqrt{a+b-2c}$ Range: 0 to N
Hamming (Manhattan or City-block) distance	$D_{AB} = \sum_{i=1}^{N} |x_{iA} - x_{iB}|$ Range: 0 to ∞	$D_{AB} = a+b-2c$ Range: 0 to N
Soergel distance	$D_{AB} = \dfrac{\sum_{i=1}^{N} |x_{iA}-x_{iB}|}{\sum_{i=1}^{N} \max(x_{iA},x_{iB})}$ Range: 0 to 1	$D_{AB} = \dfrac{a+b-2c}{a+b-c}$ Range: 0 to 1

form of the Tanimoto coefficient reduces to Equation 5.1 when the vectors consist of binary values.

It is important to note in Table 5-1 that some coefficients such as the Tanimoto and Dice coefficients measure similarity directly; others such as the Hamming and Euclidean formulae provide the *distance* (or dissimilarity) between pairs of molecules. As can also be observed from Table 5-1 some coefficients naturally provide values between zero and one; in most other cases it is possible to produce values in this range by a straightforward normalisation procedure. This provides a simple way to interconvert similarity and distance coefficients (i.e. $D = 1 - S$). The Soergel distance is the complement $(1 - S_{TAN})$ of the Tanimoto (or Jaccard) similarity coefficient when dichotomous (binary) variables are used, and was in fact developed independently of it. Our focus in this chapter is on similarity; in Chapter 6 we will discuss the selection of diverse sets of compounds for which measures of distance or dissimilarity are particularly important.

The Tversky index [Tversky 1977; Bradshaw 1997] is an example of an *asymmetric index* (i.e. $S_{AB} \neq S_{BA}$) and takes the form:

$$S_{\text{Tversky}} = \frac{c}{\alpha (a - c) + \beta (b - c) + c} \tag{5.2}$$

where α and β are user-defined constants. When $\alpha = 1$ and $\beta = 0$ then the Tversky similarity value reduces to c/a and represents the fraction of the features in A which are also in B; thus a value of one with these parameters indicates that A is a "substructure" of B. If $\alpha = \beta = 1$ then the Tversky index becomes the Tanimoto index and when $\alpha = \beta = 1/2$ it becomes the Dice coefficient.

3.1 Properties of Similarity and Distance Coefficients

For certain types of calculations it is important to know whether a distance coefficient is a *metric*. To be described as such, four properties must be met:

1. The distance values must be zero or positive, and the distance from an object to itself must be zero: $D_{AB} \geq 0$; $D_{AA} = D_{BB} = 0$
2. The distance values must be symmetric: $D_{AB} = D_{BA}$
3. The distance values must obey the triangle inequality: $D_{AB} \leq D_{AC} + D_{BC}$
4. The distance between non-identical objects must be greater than zero

The Hamming, Euclidean and Soergel distance coefficients obey all four properties; the complements of the Tanimoto, Dice and Cosine coefficients do not obey the triangle inequality (with the exception of the complement of the Tanimoto coefficient when dichotomous variables are used).

Coefficients are *monotonic* with each other if they produce the same similarity rankings. For example, the Hamming and Euclidean distances are monotonic as are the Tanimoto and Dice coefficients.

It may also be important to be aware of other characteristics of the various coefficients [Willett et al. 1998]. As can be seen from the formulae in Table 5-1, the Tanimoto, Dice and Cosine coefficients are all directly dependent upon the number of bits in common. The presence of common molecular features will therefore tend to increase the values of these coefficients. By contrast, the Hamming and Euclidean distances effectively regard a common absence of features as evidence of similarity. A consequence of this difference is that small molecules will often have lower similarity values (and larger distance separations) when using the popular Tanimoto coefficient since they naturally tend to have fewer bits set to "1" than large molecules. Smaller molecules can appear to be closer together when using the Hamming distance, which does take into account the absence of common features (i.e. values in the bitstrings count whether they are both "0" or both "1"). To illustrate some of these characteristics we show in Figure 5-3 three pairs of molecules which all differ by the substitution of a chlorine atom for a trifluoromethyl group. The Soergel and Hamming distances between these pairs of molecules are shown, calculated using the Daylight hashed fingerprints, from which some of this size dependency can be observed.

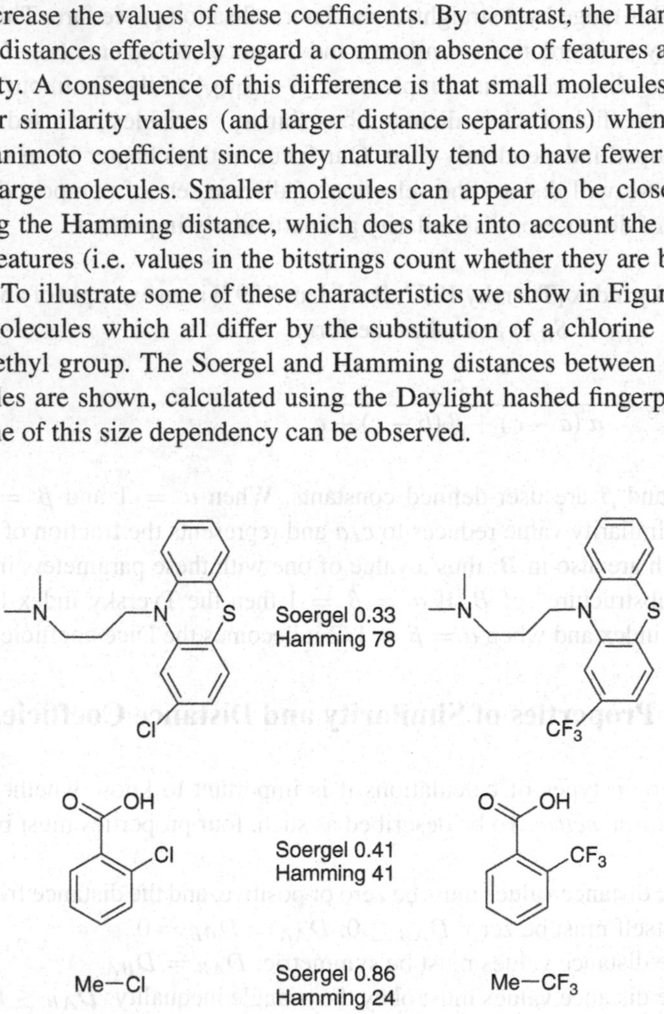

Soergel 0.33
Hamming 78

Soergel 0.41
Hamming 41

Me—Cl   Soergel 0.86   Me—CF$_3$
          Hamming 24

Figure 5-3. A comparison of the Soergel and Hamming distance values for two pairs of structures to illustrate the effect of molecular size.

One reason for the popularity of the Tanimoto coefficient for similarity searching is that it includes a degree of size normalisation via the denominator term. This can help to avoid a bias towards the larger molecules which tend to have more bits set than do smaller molecules. Conversely, the more bits set in a query molecule the higher is the average Tanimoto coefficient of the molecules in a database to it [Flower 1998]. Indeed, some workers have suggested the use of composite indices which combine two or more of the standard measures to try and overcome some of these characteristics when they cause unwanted effects to occur [Dixon and Koehler 1999].

4. OTHER 2D DESCRIPTOR METHODS

Many of the 2D descriptors described in Chapter 3 can be used to compute similarity values. Fingerprint-like descriptors such as atom pairs and topological torsions can be treated as described above. Similarity searches can also be based on continuous whole molecule properties such as the calculated $\log P$, molar refractivity and topological indices. In this case, the similarity is usually quantified using a distance coefficient such as Euclidean distance, which measures the distance between structures in a multidimensional descriptor space. Prior to calculating similarities, the data should be scaled to ensure that each descriptor makes an equal contribution to the similarity score. Statistical methods such as PCA are often used to reduce a large set of descriptors to a smaller set of orthogonal descriptors. For example, Basak et al. [1988] reduced 90 topological indices to ten principal components which were then used as the descriptors for a similarity analysis. In a similar vein, Xue et al. [1999] and colleagues found that short binary bitstrings consisting of just 32 structural fragments were more effective than conventional 2D fingerprints when combined with three numerical 2D descriptors (the number of aromatic bonds, the number of hydrogen bond acceptors and the fraction of rotatable bonds).

4.1 Maximum Common Subgraph Similarity

The similarity methods described so far do not identify any local regions of equivalence between the two structures being compared. They have therefore been called *global measures* [Downs and Willett 1995]. An alternative approach to measuring similarity involves the generation of a mapping or alignment between the molecules. An example of this is the *maximum common subgraph* (MCS). The MCS is the largest set of atoms and bonds in common between the two structures; an example is shown in Figure 5-4. The number of atoms and bonds in the MCS can be used to compute a Tanimoto-like coefficient that quantifies the

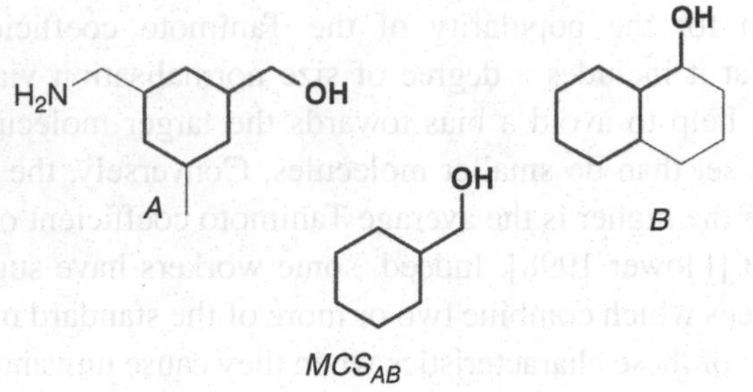

Figure 5-4. The maximum common substructure between the two molecules is shown in bold.

degree of similarity between the two compounds. For example, one could calculate the number of bonds in the MCS divided by the total number of bonds in the query structure.

Identification of the MCS in two graphs belongs to the class of problems that are NP-complete, in common with other graph searching techniques such as subgraph isomorphism. A number of both exact and approximate methods have been devised for identifying the MCS [Raymond and Willett 2002a]. The first application of the MCS to database searching was described by Hagadone [1992] who used a two-stage approach. First a fragment-based search was used to provide an upper-bound to the size of the MCS. The database was then sorted using the upper-bounds so restricting the more costly MCS calculation to those structures above a given threshold. Recently, a new algorithm called RASCAL has been described by Raymond and Willett that is able to perform tens of thousands of comparisons a minute [Raymond et al. 2002; Raymond and Willett 2002b]. In this case speed-up is achieved through the use of chemically relevant heuristics, a fast implementation of the clique detection process (see Chapter 2) that underlies MCS detection, and the use of very efficient screens that prevent many of the more costly MCS comparisons from being performed.

4.2   Reduced Graph Similarity

Various forms of reduced graph representations based on the 2D graph can be used for similarity calculations. The use of reduced graphs for the representation and searching of Markush structures in patents was described in Chapter 1. Their value in similarity searching arises from the fact that they condense a structure's key features while retaining the connections between them and so they can offer the possibility of identifying structures that have similar binding characteristics but different underlying skeletons. In addition, their smaller number of nodes makes searching operations faster. Various types of reduced graph representation have been proposed for similarity searching [Takahashi et al. 1992; Fisanick et al. 1994; Gillet et al. 2003] with the similarity between two molecules being quantified by mapping the reduced graphs to fingerprints [Gillet et al. 2003; Harper et al. 2004] and by using graph matching methods [Takahashi et al. 1992; Barker et al. 2006]. One particularly interesting type was introduced by Rarey and Dixon; when combined with fast mapping algorithms their Feature Tree representations allow similarity searching to be performed against large databases [Rarey and Dixon 1998; Rarey and Stahl 2001].

5. 3D SIMILARITY

Similarity methods that are based on rapidly calculated properties (particularly those derived from structural keys or hashed fingerprints) have become very popular, particularly when dealing with large numbers of molecules. However, methods that are based on 2D structure will tend to identify molecules with common substructures, whereas the aim is often to identify structurally different molecules.

As we have already indicated in Chapter 2, it is well known that molecular recognition depends on the 3D structure and properties (e.g. electrostatics and shape) of a molecule rather than the underlying substructure(s). An illustration of this is provided in Figure 5-5, which shows the three opioid ligands from

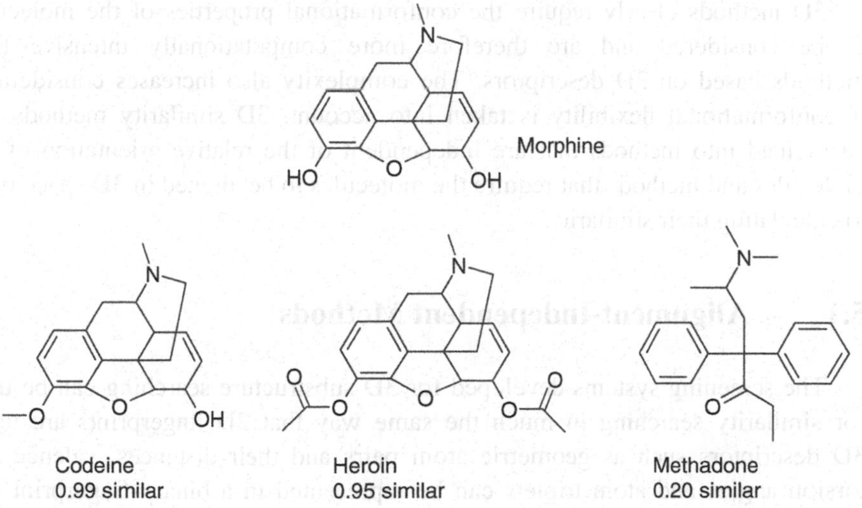

Figure 5-5. Similarities to morphine calculated using Daylight fingerprints and the Tanimoto coefficient.

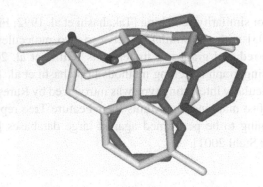

Figure 5-6. 3D overlay of morphine and methadone showing the superimposition of the basic nitrogens and the benzene rings.

Figure 5-1 together with a fourth (methadone). The similarity between morphine and the three other molecules as measured using the Daylight hashed fingerprints and the Tanimoto coefficient is shown in the figure. These similarity values reflect the clear structural similarities between morphine, codeine and heroin. Methadone, by contrast, has a very low similarity score, despite the fact that it is also active against opioid receptors. Even from the representations shown in Figure 5-5 it should be clear that it might be possible to align the conformations of the molecules to give a good overlap of the benzene rings and the nitrogen atoms. Such an overlay is shown in Figure 5-6. For reasons such as this there has been much interest in similarity measures based on 3D properties.

3D methods clearly require the conformational properties of the molecules to be considered and are therefore more computationally intensive than methods based on 2D descriptors. The complexity also increases considerably if conformational flexibility is taken into account. 3D similarity methods can be divided into methods that are independent of the relative orientation of the molecules and methods that require the molecules to be aligned in 3D space prior to calculating their similarity.

5.1   Alignment-Independent Methods

The screening systems developed for 3D substructure searching can be used for similarity searching in much the same way that 2D fingerprints are used. 3D descriptors such as geometric atom pairs and their distances, valence and torsion angles and atom triplets can be represented in a binary fingerprint and then used with the Tanimoto coefficient exactly as for 2D fingerprints [Fisanick et al. 1992]. However, to calculate all of the atom pairs and triplets in a molecule can be time consuming especially when conformational flexibility is taken into account.

Other 3D similarity methods are based on the 3D equivalent of the MCS [Moon and Howe 1990; Pepperrell and Willett 1991]. This is the largest set of atoms with matching interatomic distances (within some user-defined tolerances). As noted previously these methods are computationally expensive. When using a pharmacophore representation of the molecule then the 3D MCS problem is closely related (and in some cases identical) to the identification of the pharmacophore(s) common to a set of active compounds (see Chapter 2).

Many 3D approaches are based on the use of distance matrices where the value of each element (i, j) equals the interatomic distance between atoms i and j. One approach that aims to identify local regions of equivalence was described by Pepperrell et al. [1990]. Consider an atom i in the first molecule. The ith row of the distance matrix for molecule A contains the distances from i to every other atom. This is compared with each of the rows i' for the distance matrix for molecule B to identify the matching distances (equal within some tolerance value). A similarity between atom i from molecule A and atom i' from molecule B can be computed using the Tanimoto coefficient. This provides an *atom-match matrix* of similarities between all atoms in A and all atoms in B. This atom-match matrix can then be used to identify pairs of atoms which have geometrically similar environments, from which a global intermolecular similarity between the two molecules can then be calculated.

The pharmacophore key descriptors introduced in Chapter 3 are increasingly used for similarity comparisons. As binary representations they can be compared using the Tanimoto coefficient, analogous to 2D fingerprints. However, the characteristics of pharmacophore keys are often rather different to those of 2D fingerprints. This can affect their performance in similarity searching. In particular, pharmacophoric fingerprints tend to be more sparse than 2D fingerprints (i.e. a smaller proportion of the bits are set to "1"). A typical drug-like molecule may contain approximately 250 3-point pharmacophores out of a possible 180,000 (using seven features and 17 distance bins [Good and Lewis 1997]). As illustrated earlier, sparse fingerprints can result in rather low similarity values when using the Tanimoto coefficient. Small differences in the pharmacophoric features present can also lead to large changes in the fingerprint. These effects can be countered in a number of ways. For example, the number of distance bins can be reduced; groups of three adjacent bins can be set rather than just that corresponding to the calculated distance; and the Tanimoto coefficient can be modified to reduce its sensitivity to the total pharmacophore count [Mason et al. 1999; Good et al. 2001].

5.2   Alignment Methods

Alignment methods operate by manipulating and moving the molecules in 3D space, in contrast to the previous methods which do not require the relative orientation of the molecules to be considered. Alignment methods require the degrees of freedom related to the conformational flexibility of the molecules to be considered together with their relative orientations. Many different alignment techniques have been developed [Lemmen and Lengauer 2000].

Consider two molecules, each in a predefined conformation. When one structure is moved relative to the other the similarity measure will typically change. The goal is generally to determine the alignment of the two structures at which the similarity measure reaches a maximum. This corresponds to the optimal alignment between the two structures. Conformational flexibility may be dealt with most straightforwardly by generating a number of conformations for each molecule and then exploring all possible combinations of conformations. Alternatively, the conformations of the molecules can be varied at the same time as the relative orientations in an attempt to identify the best overlap.

As well as different ways of tackling the alignment problem, many different ways of representing molecules and calculating the similarity have been developed [Good and Richards 1998]. For example, molecules can be represented by their shapes (e.g. through consideration of the positions of atom centres or by considering molecular volume); by projecting their properties onto a surface; or by calculating molecular properties over all space or at regular intervals on a grid surrounding the molecules.

5.3   Field-Based Alignment Methods

One of the earliest methods for computing molecular similarity was proposed by Carbó et al. [1980] based on superimposed electron density maps. The similarity metric used in these calculations is essentially the same as the Cosine coefficient and is shown below:

$$C_{AB} = \frac{\int P_A P_B dv}{\left(\int P_A^2 dv\right)^{1/2} \left(\int P_B^2 dv\right)^{1/2}} \tag{5.3}$$

where P_A and P_B are the electron densities of molecules A and B. The numerator measures the degree of overlap between the two molecules and the denominator is a normalisation term. The integrals are performed over all space. A drawback of the Carbó coefficient is that it does not measure the magnitude of the fields being compared and thus the similarity is based on the degree of overlap only. Moreover, electron density is typically calculated using quantum mechanics and is therefore costly to compute. It was also found to be insufficiently discriminatory a property to use in similarity searching. It is therefore more common to use fields based on the Molecular Electrostatic Potential (MEP) together with various shape-derived properties. The properties are usually mapped to the vertices of 3D grids surrounding each molecule as in a CoMFA calculation (see Chapter 4). This approximation greatly simplifies the problem of integrating the property over all space. For example, the molecular electrostatic potential can be calculated at each vertex, $\mathbf{r}$, of a grid surrounding a molecule with N atoms, using the formula:

$$P_{\mathbf{r}} = \sum_{i=1}^{N} \frac{q_i}{|\mathbf{r} - \mathbf{R}_i|} \tag{5.4}$$

where q_i is the point charge on atom i, at position $\mathbf{R}_i$ from the grid point, $\mathbf{r}$. Steric and hydrophobic fields can be calculated in an analogous manner. Given some predefined relative orientation of two molecules, the degree of similarity of their corresponding fields can be computed by comparing corresponding grid points using a modified version of the Carbó coefficient; for a total of n grid points this is given by:

$$C_{AB} = \frac{\sum_{i=1}^{n} P_{A,\mathbf{r}} P_{B,\mathbf{r}}}{\left(\sum_{r=1}^{n} P_{A,\mathbf{r}}^2\right)^{1/2} \left(\sum_{r=1}^{n} P_{B,\mathbf{r}}^2\right)^{1/2}} \tag{5.5}$$

Other coefficients used to measure molecular similarity include the Hodgkin index [Hodgkin and Richards 1987] (which takes the magnitude of the fields into account as well as their shapes, see Table 5-1) and the Spearman rank-correlation coefficient. The latter involves calculating the Pearson correlation coefficient of the relative ranks of the data:

$$S_{AB} = 1 - \frac{6 \sum_{r=1}^{n} d_{\mathbf{r}}^2}{n^3 - n} \tag{5.6}$$

where $d_{\mathbf{r}}$ is the difference in the property ranking at point $\mathbf{r}$ of two structures and n is the total number of points over which the property is measured [Namasivayam and Dean 1986; Sanz et al. 1993].

Grid-based similarity calculations can be very time-consuming. For example, the brute-force approach to maximising the similarity between two molecules would involve rotating and translating one grid relative to the other in all possible ways and recalculating the similarity at each position. Fortunately, a number of methods have been devised for improving the efficiency of this process. In the *field-graph method*, the minimum-energy regions in the potential map are identified as nodes in a graph and clique detection is used to derive possible overlaps [Thorner et al. 1997]. A more common approach is to approximate the field distribution using a series of Gaussian functions [Kearsley and Smith 1990; Good et al. 1992; Grant and Pickup 1995; Grant et al. 1996]; these provide an analytical solution to the problem, leading to a substantial reduction in the time required to compute the overlap between a given orientation of two structures. The molecular electrostatic potential of a molecule can be approximated using a summation of two or three atom-centred Gaussian functions that take the following functional form:

$$G = \gamma . e^{-\alpha|\mathbf{r}-\mathbf{R}_i|^2} \tag{5.7}$$

These Gaussian functions can be substituted for the $1/r$ term in the electrostatic potential term in the Carbó or Hodgkin coefficient. The advantage is that the similarity calculation is broken down into a series of readily calculable exponent terms. Gaussian functions have also been used to provide more realistic representations of molecular shape rather than the traditional "hard sphere" models [Grant and Pickup 1995; Grant et al. 1996].

5.4   Gnomonic Projection Methods

An alternative to the use of "explicit" representations in Cartesian space is to use the technique of *gnomonic projection*. Here, the molecule is positioned at the centre of a sphere and its properties projected onto the surface of the sphere. The similarity between two molecules is then determined by comparing the spheres. Typically the sphere is approximated by a tessellated icosahedron or dodecahedron. For example, in the SPERM program, molecular properties are mapped onto vertices of an icosahedron by projecting vectors from the centre of the molecule onto its vertices and taking the value of the property at the point where the vector crosses the molecular surface [van Geerestein et al. 1990; Perry and van Geerestein 1992], Figure 5-7. A database structure can then be aligned to a target by rotating it about the centre relative to the target until the differences between the vertices from the two molecules is minimised. The similarity is based

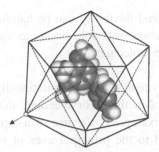

Figure 5-7. Schematic illustration of the use of gnomonic projections to compare molecules. The value at each vertex of the icosahedron is obtained by projecting a vector as indicated from the centre and determining the value of the property where the vector crosses the molecular surface. In practice each of the triangular faces of the icosahedron is divided (tessellated) into a series of smaller triangles.

on calculating the root mean squared difference between the properties calculated at each of the n vertices, $\mathbf{r}$:

$$\text{RMSD}_{AB} = \sqrt{\sum_{i=1}^{n} \frac{\left(P_{A,\mathbf{r}} - P_{B,\mathbf{r}}\right)^2}{n}} \qquad (5.8)$$

where $P_{A,\mathbf{r}}$ and $P_{B,\mathbf{r}}$ are the properties of molecules A and B at point $\mathbf{r}$. Other methods based on gnomonic projections have also been described [Chau and Dean 1987; Blaney et al. 1993; Blaney et al. 1995]. One drawback of such approaches is that the comparisons are performed only in rotational space (i.e. translations are not considered) and so the results are dependent upon the choice of the centres of projection for each molecule.

5.5 Finding the Optimal Alignment

In addition to a method for calculating molecular similarity it is generally necessary to have a mechanism for exploring the orientational (and sometimes the conformational) degrees of freedom of the molecules in order to identify the optimal alignment where the similarity is maximised. Standard minimisation methods have been used to achieve this, such the simplex algorithm [Good et al. 1992] or one of the gradient-based methods [McMahon and King 1997]. Monte Carlo methods may also be used to explore the search space [Parretti et al. 1997]. In the Field-Based Similarity Searcher (FBSS) program a GA (see Chapter 2) is used to find an optimum alignment based on steric, electrostatic or hydrophobic fields or any combination thereof [Wild and Willett 1996; Thorner et al. 1996]. Conformational flexibility can be handled directly within FBSS by allowing conformations about rotatable bonds to vary simultaneously with the relative orientations of the molecules.

3D similarity methods tend to be computationally expensive and so there is much interest in the use of methods for speeding up the calculations using some form of screening. One example is the use of a rapidly calculated volume and various properties related to the principal axes of inertia [Hahn 1997]. In this method, only those molecules that pass the screen proceed to the full alignment phase.

5.6   Comparison and Evaluation of Similarity Methods

Given the wide variety of similarity searching methods that are available it is important to be able to evaluate their performance. Since the aim of similarity searching is usually to identify molecules with similar properties, it is appropriate to test new methods on data sets for which property or activity data is available. Typically predictions are made about compounds for which the values are known. The predicted values are then compared with observed values to give an estimate of the effectiveness of a method. This is known as *simulated property prediction*. In the "leave one out" approach, each molecule is taken in turn and its nearest neighbour (i.e. the most similar molecule remaining in the data set) is found. The predicted property of the molecule is assigned to be the same as its nearest neighbour's value. Alternatively, cluster analysis (see Chapter 6) can be used to group similar compounds together, with the predicted value for the molecule being taken as the average of the values for the remaining compounds in the same cluster.

Another evaluation technique is based on the calculation of *enrichment factors* and *hit rates* [Edgar et al. 2000]. These are based on simulated screening experiments performed using databases of compounds of known activity, see Figure 5-8. Assume that a database of size N contains n_a compounds belonging to activity class a, and that all compounds not belonging to class a are inactive (or are assumed to be inactive). A simulated screening experiment involves taking an active compound as the target and ranking the entire database in order of decreasing similarity. If the similarity method is perfect then all active compounds will appear at the top of the list; if it is ineffective then the active compounds will be evenly distributed throughout the list. In practice, similarity methods usually perform somewhere between these two extremes and their effectiveness can be calculated as an enrichment factor, which is most commonly given as the ratio of the number of actives actually retrieved at a given rank compared to the number expected purely by chance. An alternative quantitive measure that takes account of

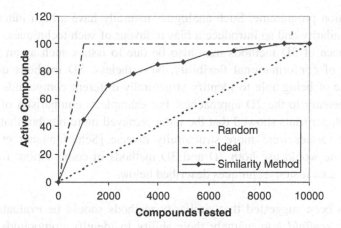

Figure 5-8. Simulated screening experiment illustrating how a similarity method can retrieve active compounds more effectively than simple random selection.

the entire ranked list is area under the receiver operator or ROC curve [Triballeau et al. 2005].

Given that molecular recognition is based on 3D properties one might expect 3D similarity methods to be more effective than 2D methods. However, this has hitherto not been found to be the case. In a celebrated and much-debated study Brown and Martin [1996] compared a range of 2D and 3D descriptors using different clustering methods and assessed their effectiveness according to how well they were able to distinguish between active and inactive compounds. They found the 2D MACCS keys to be most effective (despite the fact that, as we have mentioned previously, they were designed for optimum screenout during substructure search rather than for similarity searching). Brown and Martin [1997] also investigated the performance of different descriptors in simulated property prediction experiments. The descriptors were assessed according to their ability to predict accurately the property of a structure from the known values of other structures. The predicted properties included measured $\log P$ values together with various calculated properties that explored the shape and flexibility of the molecules. The results showed the same trend in descriptor effectiveness as before. Matter and Pötter [1999] obtained similar results from an evaluation of the ability of a wide range of 2D and 3D structural descriptors to predict biological activity.

There are several possible reasons for the apparently weaker performance of the 3D descriptors relative to the 2D descriptors. The performance of the 2D methods may be artificially enhanced due to the characteristics of the databases typically used in such evaluations. Such databases often contain large numbers of close analogues, synthesised for example as part of a lead optimisation programme. Such analogues naturally have a high inherent degree of 2D similarity and so introduce a bias in favour of such techniques. The poorer performance of 3D methods may also be due to issues such as an incomplete handling of conformational flexibility. Nevertheless, 3D methods do offer the advantage of being able to identify structurally different compounds and so are complementary to the 2D approaches. For example, a comparison of FBSS and UNITY fingerprints showed that the latter retrieved more hits but that the results from the former were more structurally diverse [Schuffenhauer et al. 2000]. Ideally, one would use both 2D and 3D methods in combination, for example, using the data fusion techniques described below.

It has been suggested that similarity methods should be evaluated on their ability to *scaffold hop*, namely their ability to identify compounds that show similar activity to the query compound but which represent different chemotypes or skeletons [Böhm et al. 2004]. Such compounds are valuable since they provide alternative lead series should one fail due to difficult chemistry or poor ADME (adsorption, distribution, metabolism and excretion) properties. Xu and Johnson [2002] have developed a scheme for classifying compounds according to their scaffolds and several groups have measured performance according to the number of different scaffolds retrieved rather than absolute number of hits [Hert et al. 2004b; Jenkins et al. 2004].

Most of the evaluations of similarity methods that have been reported in the literature are based on retrospective studies. One exception is the successful identification of novel inhibitors of the bacterial protein ZipA. The compounds were identified following similarity searches using the Rapid Overlay of Chemical Structures (ROCS) program which is based on shape. Furthermore, the compounds identified represented scaffolds that are significantly different to that of the query [Rush et al. 2005] thus demonstrating the potential of the method for scaffold-hopping applications.

A common observation when performing evaluation studies is that the effectiveness of any one similarity method can vary greatly from problem to problem and in a way that is difficult to predict. Sheridan and Kearsley [2002] have therefore advocated the use of several different methods when carrying out similarity searches. One way in which the results from different searches can be combined is to use *data fusion* techniques [Willett 2006], the expectation being that the combination of methods will improve on the results found using just one method. There are different ways in which two or more sets of data can be combined. For example, in the SUM method described by Ginn et al. [2000] the position of a molecule in hit list 1, m_1, is added to the position of the same molecule in hit list 2, m_2, to give a summed position, m_{1+2}. The new summed position is then used as the basis of a reordering of all structures to give a new fused hit list, and a new rank position for the molecule, m_{fused}. This method can be extended to any number of hit lists. Combining the hit lists from searches carried out using the same query but different similarity measures has been described as *similarity fusion* [Whittle et al. 2004]. While examples of this type of data fusion have been reported that are more effective than using just one similarity method, the performance of similarity fusion is often unpredictable. Data fusion has also been used to combine the results of searches based on different active compounds using the same similarity measure, which has been called *group fusion*. For example, extensive studies in the MDL Drug Data Report (MDDR) database [MDDR] have shown that consistent improvement in performance is achieved when using group fusion [Hert et al. 2004a], especially when the compounds used as queries are diverse. *Turbo similarity searching* is an extension of the group fusion approach which can be used when just a single active compound is available [Hert et al. 2005]. A similarity search is carried out using the method of choice. The nearest neighbours of the query are then used as seeds for new similarity searches and the resulting hit lists are combined using data fusion. Consistent with the similar property principle, the assumption is made that the nearest neighbours of the initial query compound will also be active and hence they can also be used as queries for further similarity searches. Again, extensive studies have shown that consistent improvement in performance was achieved over more conventional similarity searching which is based on the initial search only. Data fusion is also the basis of the consensus scoring approach used in protein–ligand docking, as will be discussed in Chapter 8.

6. SUMMARY

Similarity methods provide a very useful complement to techniques such as substructure and 3D pharmacophore searching. A number of similarity measures have been developed that take into account various properties of the molecules. These methods are of varying complexity, which can sometimes limit the size of the problems that can be addressed. However, even the most demanding techniques can now be used to search relatively large databases. Many studies have explored the properties of the various similarity measures with a wide variety of descriptors. The increasing number of large data sets from (HTS) will make it possible to further evaluate and test the central tenet of similarity searching, namely that structurally similar molecules have similar biological activities [Martin et al. 2002]. This is explored further in the following chapter which is concerned with the related technique of diversity analysis.

Chapter 6

SELECTING DIVERSE SETS OF COMPOUNDS

1.   INTRODUCTION

The rationale for selecting diverse subsets of compounds lies in the similar property principle introduced in Chapter 5: if structurally similar compounds are likely to exhibit similar activity then maximum coverage of the activity space should be achieved by selecting a structurally diverse set of compounds. A diverse subset of compounds or a diverse combinatorial library should be more likely to contain compounds with different activities and should also contain fewer "redundant" compounds (molecules that are structurally similar and have the same activity). In drug discovery such diverse sets of compounds may be appropriate for screening against a range of biological targets or when little is known about the therapeutic target of interest.

The widespread adoption of HTS and combinatorial chemistry techniques in the early 1990s led to a surge of interest in chemical diversity. It was widely expected that simply making "diverse" libraries would provide an increase in the number of "hits" in biological assays. However, it was soon realised that merely making large numbers of molecules was not sufficient; it was also important to take other properties into account. This in turn led to the development of concepts such as "drug-likeness" (discussed in Chapter 8) and a recognition of the need to achieve a balance between "diversity" and "focus" when designing libraries (see Chapter 9). It was also recognised that it would in any case be possible to make only an extremely small fraction of the compounds that could theoretically be synthesised. Various estimates have been made of the number of "drug-like" molecules that could be made, with figures between 10^{20} and 10^{60} being most common [Valler and Green 2000]. By contrast, the compound collection at a large pharmaceutical company may contain around 1 million molecules. Diversity methods provide a mechanism for navigating through the relevant chemical space in order to identify subsets of compounds for synthesis, purchase or testing.

As with similarity methods, there is no unambiguous definition of chemical diversity nor its quantification. As a consequence, many different ways of selecting diverse subsets have been developed. In common with similarity methods all diversity methods use molecular descriptors to define a chemistry space. A subset selection method is then required to ensure that the subset of compounds covers the defined space. It is also necessary to be able to quantify the degree of diversity of a subset of compounds in order to compare the diversity of different data sets.

The brute-force way to select the most diverse subset would be to examine each possible subset in turn. However, there are

$$\frac{N_i!}{n_i!(N_i - n_i)!} \tag{6.1}$$

different subsets of size n_i contained within a larger set of N_i compounds. For example, there are more than 10^{10} ways to select 10 compounds out of just 50. Typically, one has to deal with much larger numbers of compounds. Thus, it is computationally unfeasible to enumerate all subsets and compare them directly; approximate methods for selecting subsets must therefore be used.

This chapter concentrates on the four main approaches to selecting diverse sets of compounds: cluster analysis, dissimilarity-based methods, cell-based methods and the use of optimisation techniques.

2.   CLUSTER ANALYSIS

Cluster analysis aims to divide a group of objects into clusters so that the objects within a cluster are similar but objects taken from different clusters are dissimilar (Figure 6-1). Once a set of molecules has been clustered then a representative subset can be chosen simply by selecting one (or more) compounds from each cluster.

Cluster analysis is widely used in many diverse fields such as engineering, medicine, the social sciences and astronomy and a large number of algorithms have been devised [Aldenderfer and Blashfield 1984; Everitt 1993]. There is a wide variation in the efficiency of the various clustering methods which means that some algorithms may not be practical for handling the large data sets that are often encountered in pharmaceutical and agrochemical applications. Here we will focus on methods that are used to cluster databases of chemical structures [Willett 1987; Downs and Willett 1994; Dunbar 1997; Downs and Barnard 2002].

The key steps involved in cluster-based compound selection are as follows:

1. Generate descriptors for each compound in the data set.
2. Calculate the similarity or distance between all compounds in the data set.

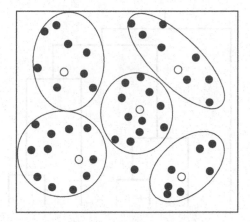

Figure 6-1. Clustering algorithms attempt to group together similar objects. A representative object might then be chosen from each cluster (open circles).

3. Use a clustering algorithm to group the compounds within the data set.
4. Select a representative subset by selecting one (or more) compounds from each cluster.

Cluster analysis is sometimes referred to as a distance-based approach to compound selection since it requires a measure of the "distance" between pairs of compounds. As explained in Chapter 5, this distance is often calculated as $1-S$ (S being the similarity coefficient) when molecules are represented by binary descriptors, and by the Euclidean distance when they are represented by physicochemical properties (or their scaled derivatives or a set of principal components).

Most clustering methods are *non-overlapping*, that is, each object belongs to just one cluster. Conversely, in *overlapping* methods an object can be present in more than one cluster. The non-overlapping methods can be divided into two classes: *hierarchical* and *non-hierarchical*.

2.1   Hierarchical Clustering

Hierarchical clustering methods organise compounds into clusters of increasing size, with small clusters of related compounds being grouped together into larger clusters: at one extreme each compound is in a separate cluster; at the other extreme all the compounds are in one single cluster [Murtaugh 1983]. The relationships between the clusters can be visualised using a *dendogram* (Figure 6-2).

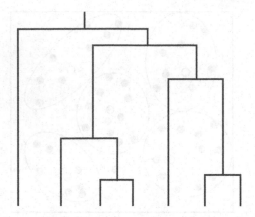

Figure 6-2. A dendrogram representing a hierarchical clustering of seven compounds.

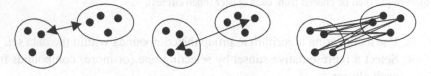

Figure 6-3. Schematic illustration of the methods used by (from left) the single linkage, complete linkage and group average approaches to calculate intercluster distances.

Agglomerative hierarchical clustering methods start at the bottom of the dendogram (all compounds in separate clusters) and proceed by merging the most similar clusters together in an iterative manner. Thus, in the first step the closest two compounds are merged into a single cluster. In the next step, the closest two clusters are merged and so on.

The most common agglomerative hierarchical methods are the so-called *Sequential Agglomerative Hierarchical Non-overlapping* (SAHN) methods. These differ in the way in which the distance or similarity between two clusters is measured.

In the *single linkage* or *nearest neighbour* method the distance between a pair of clusters is equal to the minimum distance between any two compounds, one from each cluster. The *complete linkage* or *furthest neighbour* method is the logical opposite of the single linkage method, insofar as intercluster distance is measured by the distance between the furthest pair of compounds in a pair of clusters. The *group average* method measures intercluster distance as the average of the distances between all pairs of compounds in the two clusters. These three methods are illustrated in Figure 6-3.

In *Ward's method* [Ward 1963] the clusters are formed so as to minimise the total variance. The variance of a cluster is measured as the sum of the squared deviations from the mean of the cluster. For a cluster, i, of N_i objects where each object j is represented by a vector $\mathbf{r}_{i,j}$ the mean (or centroid) of the cluster, $\bar{\mathbf{r}}_i$ is given by:

$$\bar{\mathbf{r}}_i = \frac{1}{N_i} \sum_{j=1}^{N_i} \mathbf{r}_{i,j} \tag{6.2}$$

and the intracluster variance is given by:

$$E_i = \sum_{j=1}^{N_i} \left(\left| \mathbf{r}_{i,j} - \bar{\mathbf{r}}_i \right| \right)^2 \tag{6.3}$$

The total variance is calculated as the sum of the intracluster variances for each cluster. At each iteration, that pair of clusters is chosen whose merger leads to the minimum change in total variance. Ward's method is also known as the minimum variance method.

Two additional agglomerative hierarchical clustering algorithms are the *centroid method*, which determines the distance between two clusters as the distance between their centroids, and the *median method*, which represents each cluster by the coordinates of the median value.

All six hierarchical agglomerative methods can be represented by a single equation, first proposed by Lance and Williams [1967]:

$$d_{k(i,j)} = \alpha_i d_{ki} + \alpha_j d_{kj} + \beta d_{ij} + \gamma \left| d_{ki} - d_{kj} \right| \tag{6.4}$$

$d_{k(i,j)}$ is the distance between point k and a point (i, j) formed by the merging of the points i and j. The various methods differ only in the values of the coefficients: $\alpha_i, \alpha_j, \beta$ and γ. For example, the single linkage method corresponds to $\alpha_i = \alpha_j = 0.5$, $\beta = 0$ and $\gamma = -0.5$.

Divisive hierarchical clustering algorithms start with a single cluster that contains all of the compounds (at the top of the dendogram) and progressively partition the data. A potential advantage of such an approach is that in many cases only a relatively small number of clusters is desired and so only the first part of the hierarchy needs to be produced. Divisive methods can be faster than their agglomerative counterparts, but, as reported by Rubin and Willett [1983], their performance is generally worse. This was ascribed to the fact that in the divisive methods tested the criterion for partitioning a cluster is based on a single descriptor (they are *monothetic*), in contrast to the agglomerative hierarchical methods (which are *polythetic*). They are therefore not considered further here.

2.2   Selecting the Appropriate Number of Clusters

When using an hierarchical clustering method it is necessary to choose a level from the hierarchy in order to define the number of clusters. This corresponds to drawing an imaginary line across the dendrogram; the number of vertical lines that it intersects equals the number of clusters (Figure 6-4).

One way to select the number of clusters is by examination of the dendrogram. However, several automated cluster level selection methods have also been devised. In order to compare different cluster selection methods it is necessary to be able to compare the different cluster groupings obtained. This can be done using the Jaccard statistic as shown below:

$$Jaccard(C_1, C_2) = \frac{a}{a + b + c} \tag{6.5}$$

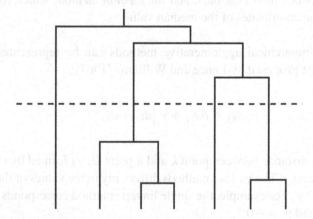

Figure 6-4. Choosing the level from the hierarchy defines the number of clusters present (in this case, four clusters).

where C_1 and C_2 are two different clusterings; a is the number of compounds that are clustered together in both clusterings; b is the number of compounds that cluster together in the first clustering but not the second; and c is the number of compounds that cluster together in the second clustering but not the first. Note that the Jaccard statistic is identical to the Tanimoto coefficient introduced in Chapter 5.

Wild and Blankley [2000] compared the effectiveness of nine different cluster selection methods using Ward's clustering applied to seven different data sets. Four different 2D binary descriptors were used: MACCS keys; Daylight fingerprints; UNITY fingerprints; and BCI fragments. The optimal clustering was chosen using each selection method and compared to the "ideal" clustering, which was established by manual clustering. The results showed that although no one approach consistently performed well across all data sets, a method proposed by Kelley et al. [1996] was found to work well with the fragment-based methods (MACCS and BCI) whereas a method termed the Variance Ratio Criterion (*VRC*) worked well with the fingerprint methods (Daylight and UNITY).

The method of Kelley et al. [1996] aims to achieve a balance between the tightness of the clusters at a particular level with the number of clusters at that level. If the number of clusters at a particular level l is k_l then the Kelley measure is given by:

$$KELLEY_l = (N - 2) \left(\frac{\langle d_{wl} \rangle - \min(\langle d_w \rangle)}{\max(\langle d_w \rangle) - \min(\langle d_w \rangle)} \right) + 1 + k_l \qquad (6.6)$$

where $\langle d_{wl} \rangle$ is the mean of the distances between objects in the same cluster at level l, and $\min(\langle d_w \rangle)$ and $\max(\langle d_w \rangle)$ are the minimum and maximum values of this property across all cluster levels. N is the total number of points in the data set. The optimal level corresponds to the smallest value of the Kelley measure.

The *VRC* compares the distances between objects within the same cluster with the distances between objects in different clusters [Calinski and Harabasz 1974]. It is defined as:

$$VRC_l = \frac{B/k_l - 1}{W/N - k_l} \qquad (6.7)$$

where B is the sum of squares of distances between objects in different clusters and W is the sum of squares of distances between objects in the same cluster. An optimal number of clusters corresponds to either a local or a global maximum in the function.

2.3   Non-Hierarchical Clustering

In non-hierarchical clustering methods compounds are placed in clusters without forming an hierarchical relationship between the clusters. A variety of non-hierarchical clustering methods exist, examples of methods used in chemical applications include the *single-pass*; *relocation*; and *nearest neighbour* methods.

As the name implies, single-pass methods cluster objects using just a single pass through the data set. The first compound encountered is assigned to the first cluster. The next compound is also assigned to this cluster if its similarity exceeds some threshold value, otherwise it is assigned to a new cluster. The process continues until all compounds have been assigned to clusters. The method is fast in operation but its main drawback is that it is order dependent; if the compounds are reordered and then reclustered a different clustering will result.

Several nearest neighbour clustering methods exist, but the method historically used for chemical applications is Jarvis–Patrick clustering [Jarvis and Patrick 1973]. In this method, the nearest neighbours of each compound are found by calculating all pairwise similarities and sorting according to this similarity. The compounds are then clustered according to the number of neighbours they have in common; two compounds are placed into the same cluster if:

1. They are in each other's list of m nearest neighbours.
2. They have p (where $p < m$) nearest neighbours in common. Typical values of m and p are 14 and 8, respectively.

A drawback of Jarvis–Patrick clustering is that it can lead to large disparate clusters and many *singletons* (clusters containing just a single molecule; of the four clusters in Figure 6-4 one is a singleton). Several modifications have been proposed to improve the effectiveness of the Jarvis–Patrick method. For example, the position of each compound within the neighbour list can be taken into account instead of simply the number of nearest neighbours. In addition, it is possible to specify that a molecule's nearest neighbours must be within some threshold distance. This ensures that the nearest neighbours of each compound are not too dissimilar.

The K-*means* method is an example of a relocation clustering method [Forgy 1965]. Here, the first step is to choose a set of c "seed" compounds. These are usually selected at random. The remaining compounds are assigned to the nearest seed to give an initial set of c clusters. The centroids of the clusters are then calculated and the objects are reassigned (or relocated) to the nearest cluster centroid. This process of calculating cluster centroids and relocating the compounds is repeated until no objects change clusters, or until a user-defined number of iterations has taken place. It is dependent upon the initial set of cluster centroids and different results will usually be found for different initial seeds [Milligan 1980].

2.4   Efficiency and Effectiveness of Clustering Methods

The efficiency of the various clustering approaches varies enormously. Some methods are not suitable for handling large data sets. All the SAHN methods can be implemented using a *stored matrix* approach in which an initial ($N \times N$) intercompound similarity matrix is first calculated and then progressively updated as new clusters are formed. However this approach requires storage (or memory) space proportional to N^2 (written O(N^2)) and the time to perform the clustering is proportional to N^3 (O(N^3)). This severely limits the applicability of these methods for clustering large data sets. A fast implementation of the Ward's method can be achieved by identifying all *Reciprocal Nearest Neighbours* (RNNs) [Murtaugh 1983]. These are pairs of objects which are mutual nearest neighbours (i.e. A is the closest object to B and *vice versa*). The RNN approach has space and time complexities of O(N) and O(N^2), respectively. The most computationally efficient of the non-hierarchical methods have a time complexity of O(MN) where M is the number of clusters generated. They are therefore very much faster than the SAHN methods. The Jarvis–Patrick method is relatively fast, having complexity O(N^2) for the calculation of the nearest neighbour lists but with a much smaller constant of proportionality than the RNN version of Ward's clustering. It has historically been used to cluster very large data sets which may be too large for any of the hierarchical methods to handle.

Various clustering methods have been compared in a number of studies. Downs et al. [1994] compared three hierarchical clustering methods (the group average and Ward's agglomerative methods and a divisive method) and the Jarvis–Patrick non-hierarchical method in a property prediction experiment. Each compound in the data set was described by 13 properties; the objective was to determine how well each of the different clustering methods was able to predict these property values. The data set was clustered using each method with the property value for each compound being predicted as the mean of the property values for the other compounds in the same cluster. The predicted values for each molecule were compared with the actual values to give a score for each cluster method. The outcome of this study was that the hierarchical methods performed significantly better than the non-hierarchical Jarvis–Patrick method in terms of their predictive ability. Interestingly, the divisive method tested (the polythetic Guenoche algorithm [Guenoche et al. 1991]) was found to be rather effective in this particular case.

Brown and Martin [1996] examined a variety of clustering methods together with several structural descriptors including structural keys, fingerprints and pharmacophore keys. The methods were evaluated according to their ability to separate a set of molecules so that active and inactive compounds were in different clusters. Four different data sets were considered. They found that a combination of 2D descriptors (particularly the structural keys) and hierarchical clustering methods were most successful. Bayada et al. [1999] also found Ward's clustering and 2D descriptors (this time using BCI fingerprints [BCI]) to be the most effective of a number of methods used to identify a subset of bioactive compounds.

Finally, it is important to bear in mind that in addition to the differences between clustering algorithms the performance of a cluster analysis also depends critically on the descriptors and similarity or distance coefficient used to calculate the distances between the objects in the data set.

3.   DISSIMILARITY-BASED SELECTION METHODS

Clustering is a two step process; first the actual clustering is performed following which a subset is chosen (e.g. by selecting the compound closest to the cluster centroid). Dissimilarity-based compound selection (DBCS) methods, by contrast, attempt to identify a diverse set of compounds directly. They are based on calculating distances or dissimilarities between compounds. The basic algorithm for DBCS is outlined below [Lajiness 1990]:

1. Select a compound and place it in the subset.
2. Calculate the dissimilarity between each compound remaining in the data set and the compounds in the subset.
3. Choose the next compound as that which is most dissimilar to the compounds in the subset.
4. If there are fewer than *n* compounds in the subset (*n* being the desired size of the final subset), return to step 2.

Steps 2 and 3 of the basic algorithm are repeated until the subset is of the required size. There are many variants on this basic algorithm; these differ in the way in which step 1 and step 3 are implemented (i.e. in the method used to choose the first compound and in the way in which the dissimilarity between each compound remaining in the data set and the compounds already placed in the subset is measured). To illustrate the procedure, the selection of five "diverse" compounds from a data set within a hypothetical 2D property space is shown in Figure 6-5.

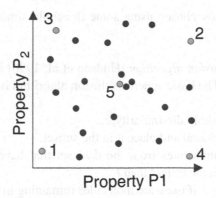

Figure 6-5. Illustration of a dissimiliarity-based compound selection in which five compounds are selected from a large set.

Three possible ways to select the initial compound are:

1. Select it at random.
2. Choose the molecule which is "most representative" (e.g. has the largest sum of similarities to the other molecules).
3. Choose the molecule which is "most dissimilar" (e.g. has the smallest sum of similarities to the other molecules).

The implementation of step 3 requires the dissimilarity values between each molecule remaining in the database and those already placed in the subset to be calculated. The most commonly used methods are MaxSum and MaxMin [Snarey et al. 1997]. If there are m molecules in the subset then the scores for a molecule i using these two measures are given by:

$$\text{MaxSum}: \text{score}_i = \sum_{j=1}^{m} D_{i,j} \qquad (6.8)$$

$$\text{MaxMin}: \text{score}_i = \text{minimum}\left(D_{i,j\,;\,j=1,m}\right) \qquad (6.9)$$

where $D_{i,j}$ is the distance between two individual molecules i and j. The molecule i that has the largest value of score_i is the one chosen. Thus MaxSum chooses the compound with the maximum sum of distances to all compounds in subset, whereas MaxMin chooses the compound with the maximum distance to its closest neighbour in the subset.

The MaxMin and MaxSum definitions are analogous to the single-linkage and group-average hierarchical agglomerative clustering methods respectively. A useful modification of these two methods is to reject any compound that is too similar to one already chosen using some threshold similarity prior to selecting the next compound.

The *sphere exclusion algorithm* [Hudson et al. 1996] is closely related to the DBCS algorithms. The basic sphere exclusion algorithm is as follows:

1. Define a threshold dissimilarity, t.
2. Select a compound and place it in the subset.
3. Remove all molecules from the data set that have a dissimilarity to the selected molecule of less than t.
4. Return to step 2 if there are molecules remaining in data set.

The first compound is selected for inclusion in the subset. Next, all compounds in the database that are within the similarity threshold to the selected compound are removed from consideration. This is analogous to enclosing the compound within a hypersphere of radius t and removing all compounds from the data set that fall within the sphere. Steps 2 and 3 are repeated until no further compounds remain in the data set. Again, several variants of the algorithm exist; these differ in the way in which the first compound is selected, in the threshold value used and in the way the "next" compound is selected in each iteration. For example, compounds can be selected at random; this is very fast in operation but the random element results in non-deterministic solutions, with different results being obtained for different runs on the same data set. Alternatively, the next compound can be selected so that it is "least dissimilar" to those already selected; Hudson et al. [1996] suggest the use of a MinMax-like method, where the molecule with the smallest maximum dissimilarity with the current subset is selected.

The relationship between the DBCS methods and sphere-exclusion algorithms is explored in the Optimizable K-Dissimilarity Selection (OptiSim) program [Clark 1997; Clark and Langton 1998]. OptiSim makes use of an intermediate pool of molecules, here called *Sample*. In order to identify the next compound to add to the final subset, molecules are chosen at random from the data set and added to *Sample* if they have a dissimilarity greater than t from those already in the final subset; otherwise they are discarded. When the number of molecules in *Sample* reaches some user-defined number, K, then the "best" molecule from *Sample* is added to the final subset (the "best" molecule being that which is most dissimilar to those already in the final subset). The remaining $K - 1$ compounds from *Sample* are then set aside but may be reconsidered should the main set run out before the subset reaches the desired size. The process then repeats. The characteristics of the final subset of selected molecules are determined by the value of K chosen, with values of K equal to 1 and to N corresponding to sphere-exclusion and DBCS respectively.

3.1 Efficiency and Effectiveness of DBCS Methods

The basic DBCS algorithm has an expected time complexity of $O(n^2N)$, where n is the number of compounds that one wishes to select from N. As n is generally a small fraction of N (such as 1% or 5%) this gives a running time that is cubic in N, which makes it extremely demanding of computational resources if the data set is large. Holliday et al. [1995] described a fast implementation of the MaxSum algorithm of complexity $O(nN)$ that enables the dissimilarity between a compound and a subset to be computed in a single operation rather than having to loop over all molecules currently in the subset.

The MaxMin dissimilarity definition is not normally as fast in practice as MaxSum, but it too can be implemented with time complexity of $O(nN)$. In fact, since MaxMin is based on identifying nearest neighbours, it is possible to use the extensive work that has been carried out on fast algorithms for nearest neighbour searching in low-dimensional spaces. Agrafiotis and Lobanov [1999] have used one such approach, based on k–d trees, to obtain a MaxMin algorithm with a time complexity of only $O(n \log N)$.

A schematic illustration of the behaviour of the MaxSum, MaxMin and sphere exclusion methods is given in Figure 6-6 for a 2D descriptor space. If the "most dissimilar" compound is chosen as the first molecule in the DBCS methods then the MaxSum method tends to select closely related compounds at the extremities of the distribution [Agrafiotis and Lobanov 1999; Mount et al. 1999; Snarey et al.

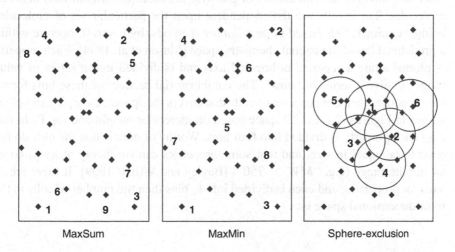

| MaxSum | MaxMin | Sphere-exclusion |

Figure 6-6. Schematic comparison of various dissimilarity selection methods. (Redrawn from Leach 2001.)

1997]. This is also the initial behaviour of the MaxMin approach which then starts to sample from the middle. Sphere exclusion methods typically start somewhere in the middle of the distribution and work outwards. The DBCS results are typically more diverse than those resulting from the sphere-exclusion approach, the latter being more likely to give a subset that is representative of the original data set.

In their work Snarey et al. [1997] examined the extent to which the methods were able to identify subsets of compounds from the WDI [WDI] with a wide range of biological activities. This database includes compounds from many different classes of biological activity (e.g. antibiotics, antihistamines, analgesics); the various subsets selected were scored according to the number of distinct activity classes present. The more activity classes selected the "better" the performance of the algorithm. The molecules were represented by hashed fingerprints or by a series of topological indices and physical properties and the dissimilarities, (D_{ij} in equations 6.8 and 6.9) were measured by the complement of either the Tanimoto or the cosine coefficient. The results suggested that there was relatively little to choose between the best of these methods (though some proved significantly worse than a random selection) but the MaxMin maximum-dissimilarity algorithm was generally considered to be both effective and efficient.

4.   CELL-BASED METHODS

Clustering and DBCS are sometimes referred to as distance-based methods since they involve the calculation of pairwise distances (or similarities) between molecules. The results are thus dependent upon the particular set of molecules being examined. *Cell-based* or *partitioning* methods, by contrast, operate within a predefined low-dimensional chemistry space [Mason et al. 1994]. Each property is plotted along a separate orthogonal axis and is divided into a series of value ranges to give a series of "bins". The combinatorial product of these bins for all the properties then defines a set of cells that covers the space. A simple example is shown in Figure 6-7 for a 2D space based on molecular weight and log P. In this case each property is divided into four bins. Worthy of note is that the bins do not need to be equal in size, and that some properties can (in theory at least) cover an infinite range (e.g. "$MW > 750$") [Bayley and Willett 1999]. If there are N axes, or properties, and each is divided into b_i bins then the number of cells in the multidimensional space is:

$$\text{Number of cells} = \prod_{i=1}^{N} b_i \qquad (6.10)$$

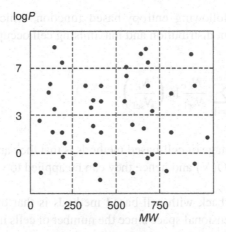

Figure 6-7. The construction of a 2D chemistry space. In this case the log P bins are <0, $0–3$, $3–7$ and >7 and the MW bins are $0–250$, $250–500$, $500–750$ and >750.

Compounds are allocated to cells according to their molecular properties. Partitioning is a two-stage process, like clustering; once compounds have been allocated to the cells then a representative subset can be selected by choosing one or more compounds from each cell.

A key feature of cell-based methods is that they do not require the calculation of pairwise distances between the compounds; the chemistry space is defined independently of the molecules that are positioned within it. This offers two key advantages. First, empty cells (*voids*) or cells with low occupancy can be readily identified; these indicate regions of the space that are under-represented. Second, the diversity of different subsets can be easily compared by examining the overlap in the cells occupied by each subset. Neither of these features is straightforward to achieve using clustering or DBCS. An alternative to the simple diversity measure that involves counting the number of occupied cells is the following χ^2 statistic [Waldman et al. 2000]:

$$D_{\chi^2} = - \sum_{i=1}^{N_{\text{cells}}} \left(N_i - \frac{N_{\text{sel}}}{N_{\text{cells}}} \right)^2 \tag{6.11}$$

where N_i is the number of molecules in cell i, N_{sel} is the number of molecules in the selected subset and N_{cells} is the total number of cells in the space. This metric tends to favour a uniform distribution of molecules throughout the space. A third alternative is the following entropy-based function, which achieves a balance between the uniform distribution and maximising cell occupancies:

$$D_{\text{entropy}} = -\sum_{i=1}^{N_{\text{cells}}} \frac{N_i}{N_{\text{sel}}} \ln\left(\frac{N_i}{N_{\text{sel}}}\right) \tag{6.12}$$

A further advantage of cell-based methods is that they are very fast with a time complexity of just $O(N)$ and hence they can be applied to very large data sets.

The main drawback with cell-based methods is that they are restricted to a relatively low-dimensional space since the number of cells increases exponentially with the number of dimensions N. It is therefore necessary to ensure that a small number of relevant descriptors are selected. Thus it is inappropriate to base the axes on the bits in a bitstring, such as a structural key or hashed fingerprint; for a bitstring of length n the number of cells would be 2^n. For typical bitstrings this would give an impossibly large number of cells (e.g. 2^{1024} for a 1,024-bit bitstring), almost all of which would be empty. Also the cell boundaries are somewhat arbitrary and moving the boundaries, even by a small amount, can have a significant impact on the compounds selected.

One of the first approaches to partitioning was described by Lewis et al. [1997]. The aim was to develop a partitioning scheme based on physiochemical properties considered important for both drug-receptor binding and for the transport of a drug to its target. Following a statistical analysis of nearly 50 descriptors, they identified six that were weakly correlated and that covered properties such as hydrophobicity, polarity, shape, hydrogen bonding properties and aromatic interactions. Each descriptor was split into two, three or four partitions giving a grand total of 576 cells. When a subset of the corporate database consisting of 150,000 molecules was mapped onto the space 86% of the defined cells were occupied. Some of the cells were not occupied since they represent combinations of properties that are unlikely to exist, for example, a hydrophobic molecule having a large number of hydrogen bonding groups. A representative set of compounds designed to act as a small diverse screening set was then identified by choosing three compounds from each occupied cell.

The BCUT descriptors were developed by Pearlman and Smith [1998] for the purposes of providing a low-dimensional chemistry space. As described in Chapter 3, BCUTs are based on matrix representations of molecules whereby atomic properties such as atomic charges, polarisabilities and hydrogen bonding capabilities are used for the diagonal elements and various connectivity representations are used for the off-diagonal elements. Consideration of three atomic properties gives rise to three different matrices with the highest and lowest eigenvalues providing six descriptors to form the axes of the partitioned space. However, as indicated in Chapter 3, many variants on these matrices are possible. For example, the off-diagonal elements can be derived from the bonding information in the 2D graph or from interatomic distances from a 3D conformation. Moreover, different methods can be used to calculate the diagonal elements (e.g. alternative methods to calculate atomic charges). This can result in a large number of potential choices for the axes of the partitioned chemistry space. Pearlman and Smith [1999] have therefore developed methods for selecting the most appropriate set of axes, to ensure that the compounds are as uniformly distributed in the space as possible.

A high-dimensional descriptor space can be reduced to a low-dimensional space appropriate for partitioning by using a technique such as PCA (discussed in Chapter 3). This results in a small set of orthogonal axes with each axis corresponding to a linear combination of the original descriptors. Such an approach was used in a comparative study of various chemical databases [Cummins et al. 1996]. The initial descriptors comprised the computed free energy of solvation of each molecule together with a large number of topological indices. Following the removal of highly correlated descriptors and descriptors that showed little variation, factor analysis was applied. This is a technique similar in spirit to PCA and it resulted in four factors that explained 90% of the variation in the data. The 4D space was partitioned into cells and molecules from five databases were projected into the space. Most of the molecules were found to occupy a relatively small region of the space and so an iterative procedure was used to remove outliers, thereby enabling the resolution to be increased in the area populated by the majority of the molecules. The databases were then compared two at a time by counting how many cells they had in common. It was of particular interest to identify the regions of space occupied by those databases that contain solely biologically active molecules, and to see whether this space was different to that occupied by the more general databases.

4.1 Partitioning Using Pharmacophore Keys

3D-pharmacophore descriptors can be used to provide a partitioned space. Here, each potential 3- or 4-point pharmacophore is considered to constitute a single cell. A molecule can be mapped onto the space by identifying the 3- or 4-point pharmacophores that it contains. Unlike the partitioning schemes described above that are based on whole molecule properties in which each molecule occupies a single cell, in pharmacophore partitioning a molecule will typically occupy more than one cell since almost all molecules will contain more than one

3- or 4-point pharmacophore. The pharmacophore keys for a set of molecules can also be combined into an ensemble pharmacophore which is the union of the individual keys. The ensemble can then be used to measure total pharmacophore coverage to identify pharmacophores that are not represented and to compare different sets of molecules.

One of the earliest programs to use pharmacophore keys to select a diverse set of compounds was the simple order-dependent algorithm implemented within the Chem-Diverse software [Davies 1996]. The goal was to select subsets of compounds that would maximise the total pharmacophore coverage. The pharmacophore key is generated for the first molecule in the database. This becomes the ensemble pharmacophore. The pharmacophore key for the next molecule in the list is then generated and compared with the ensemble. If the new molecule covers some amount of pharmacophore space not yet represented in the ensemble, then it is added to the subset and the ensemble is updated, otherwise the molecule is rejected. The procedure continues until sufficient molecules have been generated or the ensemble pharmacophore has become saturated.

This approach does suffer from some limitations. For example, the ensemble only records the presence or absence of a particular pharmacophore and does not take into account the frequency of occurrence. No account is taken of promiscuous molecules (molecules that contain a large number of pharmacophores, often very flexible molecules). Subsequent approaches have addressed these problems and have also enabled additional properties to be taken into account when selecting a subset or designing libraries [Good and Lewis 1997].

5.   OPTIMISATION METHODS

As indicated in the introduction, only for the smallest of systems is it possible to contemplate a systematic evaluation of all possible subsets. Many of the methods that we have discussed so far for selecting a diverse subset involve procedures that iteratively select one molecule at a time. An alternative is use some form of optimisation procedure. Optimisation techniques provide effective ways of sampling large search spaces and several of these methods have been applied to compound selection. One of the first published methods was that of Martin et al. [1995] who used D-optimal design. As outlined in Chapter 4 this is a statistical experimental design technique that aims to produce a subset of compounds that is evenly spread in property space. However, the technique is somewhat prone to selecting outliers as well as molecules that are very similar to each other.

Many approaches are based on Monte Carlo search, often with simulated annealing [Hassan et al. 1996; Agrafiotis 1997]. An initial subset is chosen at random and its diversity is calculated. A new subset is then generated from the first by replacing some of the compounds with others chosen at random from the data set. The diversity of the new subset is measured; if this leads to an increase in the diversity function then the new subset is accepted for use in the next iteration. If the new subset is less diverse, then the probability that it is accepted depends on the Metropolis condition, $\exp[-\Delta E/k_B T]$, where ΔE is the difference between the current and previous diversity values (see Chapter 2 for more details). The process continues for a fixed number of iterations or until no further improvement is observed in the diversity function. In the simulated annealing variant the temperature of the system is gradually reduced, so increasing the chance of finding the globally optimal solution.

Typical diversity functions used in such procedures include MaxMin and MaxSum; however, there continues to be interest in the development of new measures of molecular diversity. Introducing a new function, Waldman et al. [2000] suggested five requirements that a diversity function should possess:

1. The addition of a redundant molecule (i.e. one that has the same set of the descriptors as an existing molecule in the set) should not change the diversity.
2. The addition of non-redundant molecules should always increase the diversity.
3. The function should favour space-filling behaviour, with larger increases in the function for filling large voids in the space rather than filling more heavily populated regions.
4. If the descriptor space is finite, then filling it with an infinite number of molecules should give a finite value for the diversity function.
5. As one molecule is moved further away from the others then the diversity should increase, but it should asymptotically approach a constant value.

The approach proposed by Waldman was based on the overlap of Gaussian functions. The overlap is calculated by computing the *minimum spanning tree* for the set of molecules. A spanning tree is a set of edges that connect a set of objects without forming any cycles. The objects in this method are the molecules in the subset and each edge is labelled by the dissimilarity between the two molecules it connects. A minimum spanning tree is the spanning tree that connects all molecules in the subset with the minimum sum of pairwise dissimilarities, Figure 6-8. The diversity then equals the sum of the intermolecular similarities along the edges in the minimum spanning tree.

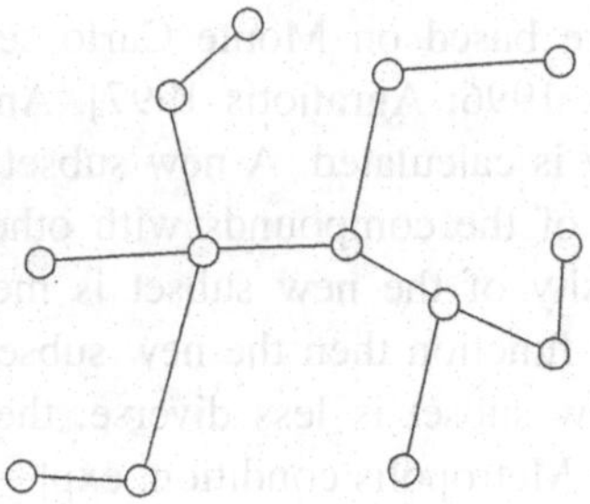

Figure 6-8. Illustration of a minimum spanning tree for a simple 2D system.

The computational requirements of the optimisation methods are dependent on the diversity function used since the function is typically applied very many times during the optimisation process. Thus, efficient operation of these methods requires diversity functions that are rapid to calculate.

6.   COMPARISON AND EVALUATION OF SELECTION METHODS

Given the different compound selection methods and the many different descriptors that are available it is natural to consider which methods are most appropriate [Willett 1999]. The aim when selecting diverse subsets is to minimise the number of compounds that share the same activity whilst maximising the coverage of different bioactivity types. Several studies have attempted to evaluate the various "rational" approaches to compound selection according to their ability to extract compounds with known activities from within larger data sets.

Some of these studies have suggested that the rational methods are no better than random at selecting bioactive molecules. For example, Taylor [1995] compared cluster and dissimilarity-based selection in a simulation experiment and concluded that cluster-based selection was only marginally better than random and that dissimilarity-based selection was worse than random. However, other studies have found the converse. For example, Lajiness [1991] determined that both cluster-based and dissimilarity-based compound selection methods were better than random. Bayada et al. [1999] compared the MaxMin DBCS with Ward's clustering, partitioning and a Kohonen mapping method (Kohonen maps are discussed in Chapter 7) and found that Ward's clustering was superior to the other methods. Pötter and Matter [1998] compared DBCS and hierarchical clustering with random selection methods and found the DBCS methods to be more effective than the random selections. The comparison of DBCS and sphere exclusion methods by Snarey et al. [1997] has already been mentioned.

The selection of diverse subsets is based on the premise that structural similarity is related to similarity in biological activity space. Brown and Martin [1996] in their clustering experiments concluded that compounds within 0.85 Tanimoto similarity (calculated using UNITY fingerprints) have an 80% chance of sharing the same activity. These results were used to guide the selection of compounds from external suppliers such that no molecule within 0.85 similarity of any compound already present in the screening set was purchased. In a comparison of 2D and 3D descriptors Matter and Pötter [1999] also confirmed the 0.85 similarity threshold for UNITY fingerprints and the Tanimoto coefficient. However, recent work by Martin et al. [2002] has shown that the relationship between structural similarity and biological activity similarity may not be so strong as these previous studies suggested. They found that a molecule within 0.85 similarity of an active compound had only a 30% chance of also being active. Nevertheless, the enrichments achieved were still significantly better than random. Thus, it was suggested that a screening library should contain some number of similar compounds while still maintaining its diversity.

7.   SUMMARY

The four main approaches to selecting diverse subsets of compounds have been described; clustering, dissimilarity-based methods, cell-based methods and optimisation techniques. To date, there is no clear consensus as to which method is the best. When choosing which method or methods to apply several criteria should be taken into account. First, it should be noted that the choice of subset selection method and the descriptors employed are interrelated. Some subset selection methods can be used with high-dimensionality descriptors such as binary fingerprints (e.g. clustering and DBCS methods) whereas cell-based methods can only be used with low-dimensionality data, or with data that has been subjected to some form of dimensionality reduction. Second, the relative efficiencies of the various approaches is a factor for consideration and the more computationally demanding methods such as clustering may be restricted to smaller data sets (though fast implementation methods and the introduction of parallel clustering algorithms means that this is becoming less of a restriction). The computational requirements of the optimisation methods are typically determined by the speed of the fitness function, since this is calculated very frequently throughout the search. Finally, it is important that the descriptors used are relevant to biological activity; some pre-analysis of the data should be carried out in order to ensure that an appropriate set of descriptors is chosen.

Chapter 7

ANALYSIS OF HIGH-THROUGHPUT SCREENING DATA

1.   INTRODUCTION

Robotic and miniaturisation techniques now make it possible to screen large numbers of compounds very efficiently. Current HTS systems enable hundreds of thousands of compounds to be assayed in only a few days [Hertzberg and Pope 2000]. HTS runs can generate immense volumes of data for analysis. In most current systems the data obtained from HTS comprises "single-shot" percentage inhibition data derived by measuring the activity of each sample at one predetermined concentration. The data set is then analysed to identify a subset of compounds that are progressed to the next stage, in which a more comprehensive measure of the inhibition is determined. A common procedure is to determine a dose–response curve by measuring activity at different concentrations, from which an IC50 value can be determined (the IC50 is the concentration of inhibitor required to reduce the binding of a ligand or the rate of reaction by 50%). Depending on the numbers involved, these assays may be performed using pre-dissolved liquid samples; to provide final confirmation a dose–response curve is determined from a solid sample whose purity is also checked. If the sample is pure and the compound is deemed to be of interest then elaboration of the structure using chemical synthesis may commence (often referred to as the "hits-to-leads" phase). Analysis of the data resulting from one HTS run can also be used to design new compounds to screen in a subsequent iteration of the drug discovery cycle. The entire process is illustrated in Figure 7-1; key to our discussion is that at each stage it is necessary to analyse the relevant data and to select which compounds to progress to the next stage.

The simplest way to deal with the output from HTS is to select the most potent compounds to progress to the next stage, the number selected being determined by the throughput of the subsequent assay. However, such an approach may not necessarily result in the most useful structures being selected for progression. Some molecules may contain functional groups that cause interference in the assay (e.g. fluorescent compounds); other molecules may contain functional groups that are known to react with biological systems. Both these situations may give rise to

141

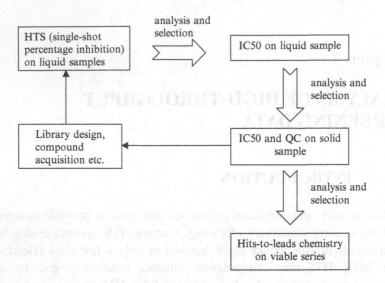

Figure 7-1. A schematic illustration of the process used to progress the initial hits from an HTS run through to synthetic chemistry. QC stands for Quality Control, a procedure to determine the purity and integrity of the sample using one or more analytical techniques.

misleading data values. However, such compounds can often be readily identified and eliminated if necessary using simple substructure and "drug-likeness" filters; these will be considered in more detail in Chapter 8. Other compounds may simply be unattractive starting points for chemical elaboration. There can be significant noise in the biological data which means that, particularly for single-shot data, the measured activity may not be an accurate quantitative indication of the true biological activity. In some cases individual compounds may appear to be particularly potent but closely related analogues may have no activity. Some of these scenarios are illustrated in Figure 7-2. For these and other reasons it is therefore desirable to be able to analyse the data from HTS in order to identify the best samples to progress, with the ultimate goal being to identify as many potential lead series as possible.

It might be considered that the statistical methods described in Chapter 4 could be applied to the analysis of HTS data. Unfortunately, the nature of the data means that methods such as multiple linear regression or PLS are not applicable. Such methods work best with small data sets containing high-quality data, and where the compounds are from the same chemical series (or at least are closely related to each other and operate by the same biological mechanism). HTS data by contrast is characterised by its high volume, by its greater level of noise, by the diverse nature of the chemical classes involved and by the possible presence of multiple binding modes.

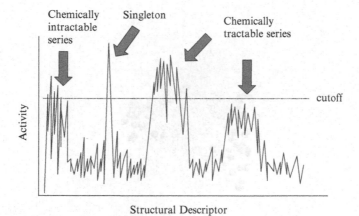

Figure 7-2. Schematic illustration of the way in which the biological activity measured in an HTS run may vary with the chemical structure. The diagram shows a potent but chemically intractable series (in addition, this set of compounds shows greater fluctuations in activity), a highly potent but isolated compound (singleton), and two tractable series, one of which may be missed by the application of a simple activity cut-off.

The remainder of this chapter will be concerned with a description of some of the methods that can be used for the analysis of HTS data [Gedeck and Willett 2001; Bajorath 2002]. The focus will be on methods that are applicable to large, structurally diverse sets of compounds. This includes data visualisation and data reduction techniques. When activity data is available then one may wish to determine relationships between the chemical structure and the observed activity using data mining techniques. It should also be noted that most of the techniques discussed here can also be applied to smaller data sets.

2. DATA VISUALISATION

One of the most important aspects of data analysis is visualisation. Molecular graphics packages for the display of the results of computational chemistry calculations have been available for many years. However, most of these are concerned with the display of the 3D structures and conformationally dependent properties of single systems (albeit very complex ones such as protein–ligand complexes). By contrast, the need in chemoinformatics is often for the simultaneous display of large data sets containing many thousands of molecules and their properties. Several packages are now available for the graphical display of large data sets; the facilities provided by such packages typically include the capability to draw various kinds of graphs, to colour according to selected properties and to calculate simple statistics. Some of this software is

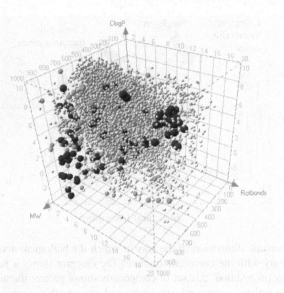

Figure 7-3. Graphical representation of the property distributions of active (large spheres), moderately active (medium spheres) and inactive (small spheres) compounds from the NCI AIDS data set [NCI] using Spotfire [Spotfire; Ahlberg 1999].

designed specifically for the chemical domain; other programs are applicable to many different types of data. By way of illustration, Figure 7-3 shows a set of approximately 300 active and 900 moderately active compounds from the National Cancer Institute (NCI) AIDS antiviral screen [NCI], together with about 38,000 inactive structures. These have been plotted according to three properties (molecular weight, ClogP and the number of rotatable bonds). One feature that is often very useful for the analysis of chemical data sets is the ability to display and manipulate selected chemical structures (e.g. to interactively perform a substructure search to identify all compounds containing a particular feature).

HTS data sets are often so large that it may help to divide the molecules into subsets in order to help navigation through the data. This can be achieved either using unsupervised methods that do not use the activity data or supervised methods that do use the activity data. In this section we will consider methods in the former category; supervised approaches are discussed in Section 3.

Perhaps the simplest strategy is to use a technique such as cluster analysis to divide the structures into groups of structurally similar molecules or molecules that have similar properties. One may then, for example, calculate the mean activity of each cluster and use this to identify any clusters where the mean activity is significantly above the average activity for the entire data set for further investigation. The cell-based methods described in Chapter 6 may also be used; certain cells of the chemical space may contain a greater proportion of active molecules. Another approach is to use predefined sets of substructures which then form the basis for data visualisation and navigation [Roberts et al. 2000]. By choosing substructures that are "meaningful" to a medicinal chemist it may be possible to rapidly identify specific classes of molecules to investigate further.

2.1 Non-Linear Mapping

It is common practice to compute molecular descriptors that might be useful for understanding and explaining the relationship between the chemical structure and the observed activity. These descriptors may include various physicochemical properties, fingerprints, the presence of substructural fragments, and any known activity against related targets. HTS data sets are thus usually multidimensional in nature. In order to visualise a multidimensional data set it often helps to map it to a lower 2D or 3D space. This process is usually referred to as *non-linear mapping* the objective of which is to reproduce the distances in the higher-dimensional space in the low-dimensional one. It is often particularly desirable that objects close together in the high-dimensional space are close together in the low-dimensional space. Having reduced the data set to 2D or 3D one may then use graphics or other visual cues to identify structure–activity relationships, for example by colouring the data points according to the biological activity.

A commonly used procedure for non-linear mapping is *multidimensional scaling* [Kruskal 1964; Cox and Cox 1994]. There are two stages involved. First, an initial set of coordinates is generated in the low-dimensional space. This may be achieved in a number of ways, for example by using principal components analysis or simply by generating a set of random coordinates. In the second phase these coordinates are modified using a mathematical optimisation procedure that improves the correspondence between the distances in the low-dimensional space and the original multidimensional space. To achieve this the optimisation procedure uses an objective function, a common function being *Kruskal's stress*:

$$S = \frac{\sum_{i<j}^{N} \left(d_{ij} - D_{ij}\right)^2}{\sum_{i<j}^{N} d_{ij}^2} \tag{7.1}$$

where d_{ij} is the distance between the two objects i and j in the low-dimensional space and D_{ij} is the corresponding distance in the original multidimensional space. The summations are over all pairs of distances for the N points in the data set. The optimisation continues until the value of the stress function falls below a threshold value.

An alternative function used in *Sammon mapping* [Sammon 1969] is the following:

$$S = \frac{\sum\limits_{i<j}^{N} \dfrac{(d_{ij} - D_{ij})^2}{D_{ij}}}{\sum\limits_{i<j}^{N} D_{ij}} \tag{7.2}$$

The key difference between the Kruskal and the Sammon functions is that the latter places more emphasis on the smaller distances due to the presence of the normalisation term. This can be observed in the simple example shown in Figure 7-4. The main drawback of the two methods as originally devised is that

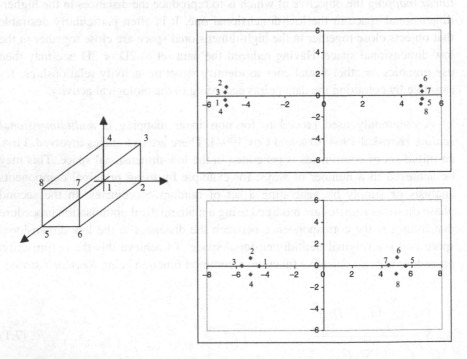

Figure 7-4. Simple illustration of the difference between Kruskal and Sammon functions for non-linear mapping. A rectangular box with sides of length 1, 1 and 10 (left) was mapped into 2D using both the Kruskal (top graph) and Sammon (bottom graph) stress functions. As can be seen, the Sammon function tends to preserve the shorter distances better than the Kruskal function.

they scale with the square of the number of points, making their application to large data sets problematic. This is primarily due to the limitations of the optimisation procedure. Several solutions that overcome this limitation have been devised [Xie et al. 2000; Agrafiotis and Lobanov 2000], making it possible to apply these techniques to data sets containing tens of thousands of data points.

3. DATA MINING METHODS

Data mining techniques are now widely used to identify relationships in large, multidimensional data sets in many areas and the analysis of HTS data is no exception. A key objective of such an analysis is the construction of models that enable relationships to be identified between the chemical structure and the observed activity. As indicated above, traditional QSAR methods such as multiple linear regression are not generally applicable and so alternative techniques have been investigated. It is often more appropriate for HTS data sets to classify the molecules as "active" or "inactive" or into a small number of activity classes (e.g. "high", "medium", "low") rather than using the numerical activity. The aim is to derive a computational model that enables the activity class of new structures to be predicted. For example, the model could be used to select additional compounds for testing, to design combinatorial libraries or to select compounds for acquisition from external vendors.

3.1 Substructural Analysis

Substructural analysis (SSA) [Cramer et al. 1974] is related to the Free–Wilson approach described in Chapter 4. The premise is that each substructural fragment makes a constant contribution to the activity, independent of the other fragments in the molecule. The aim is to derive a weight for each substructural fragment that reflects its tendency to be in an active or an inactive molecule. The sum of the weights for all of the fragments contained within a molecule gives the score for the molecule, which therefore enables a new set of structures to be ranked in decreasing probability of activity.

Many different weighting schemes are possible. One fairly simple function is to define the weight of a fragment i according to:

$$w_i = \frac{\mathrm{act}_i}{\mathrm{act}_i + \mathrm{inact}_i} \qquad (7.3)$$

where act_i is the number of active molecules that contain the ith fragment and $inact_i$ is the number of inactive molecules that contain the ith fragment. An early application of SSA was to screening set selection in the US Government's anti-cancer programme [Hodes et al. 1977]. *Naive Bayesian classifiers* (NBCs) are closely related to SSA [Hert et al. 2006] and have also been applied to screening set selection and the analysis of HTS data sets [Bender et al. 2004; Xia et al. 2004; Rogers et al. 2005].

The fragments used in SSA often correspond to those present in structural keys of the type used for substructure searching. An alternative is to use an automated method to identify substructures that are relevant for a given data set. This is the basis of the CASE and MULTICASE algorithms [Klopman 1984; Klopman 1992]. These methods generate all possible connected fragments, containing between 3 and 12 atoms, for the molecules in the data set. The distribution of each fragment in the active and inactive molecules is determined. Each distribution is then evaluated to determine whether the fragment is particularly prevalent in the active molecules or in the inactive molecules or whether its distribution is simply that which would be expected by chance. In this way it is possible to automatically identify a set of fragments specific to the data set that constitute a mathematical model for predicting whether new structures are likely to be active or inactive. The CASE and MULTICASE algorithms have been applied to a wide variety of problems including the prediction of physicochemical properties [Klopman et al. 1992; Klopman and Zhu 2001], the generation of quantitative structure–activity models (particularly for predicting toxicity, carcinogenicity, mutagenicity and biodegradation [Rosenkranz et al. 1999]) and to the analysis of large data sets [Klopman and Tu 1999]. It is also possible to use hashed fingerprints for substructural analyses. For example, in the "hologram QSAR" approach PLS analysis is used to derive a model based on the number of times each bit in the fingerprint is set [Tong et al. 1998]. A further example is provided by the Stigmata algorithm which identifies bits that are present in a user-defined fraction of the active molecules with the results being presented using graphical display techniques [Shemetulskis et al. 1996].

3.2   Discriminant Analysis

The aim of *discriminant analysis* is to try and separate the molecules into their constituent classes. The simplest type is *linear discriminant analysis* which in the case of two activity classes and two variables aims to find a straight line that best separates the data such that the maximum number of compounds is correctly classified [McFarland and Gains 1990] (see Figure 7-5). With more than two variables the line becomes a hyperplane in the multidimensional variable space.

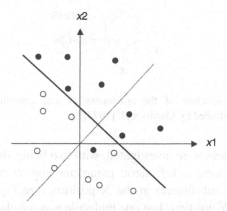

Figure 7-5. A linear discriminant analysis aims to find a line that best separates the active and the inactive molecules. The dotted line in the figure is the discriminant function; the solid line is the corresponding discriminant surface. Note that in this case it is not possible to find a line that completely separates the two types of data points.

A linear discriminant analysis is characterised by a discriminant function which is a linear combination of the independent variables $(x_1, x_2, \ldots, x_n)$:

$$W = c_1 x_1 + c_2 x_2 + \cdots c_n x_n \tag{7.4}$$

The surface that separates the classes is orthogonal to this function, as shown in Figure 7-5. Feeding the appropriate descriptor values into this equation enables the value of the discriminant function to be computed for a new molecule. Values above a threshold correspond to one activity class and lie to one side of the discriminant surface and values below the threshold correspond to the other activity class. More than one discriminant function may be possible, corresponding to different values for the coefficients c_i in Equation 7.4 (i.e. there is not necessarily a unique solution). Techniques such as cross-validation may be useful in order to distinguish between the different functions.

The first application of discriminant analysis to a set of biologically active molecules is generally considered to be the work of Martin et al. [1974] on a series of aminotetralins and aminoindans that are inhibitors of the enzyme Monoamine Oxidase, a potential antidepressant target. The study involved a relatively small data set containing 20 compounds with the general structure shown in Figure 7-6. Linear discriminant analysis was used for the analysis rather than multiple linear regression since the biological activity data were not on a continuous scale of activity but rather were classified into two groups: active and inactive.

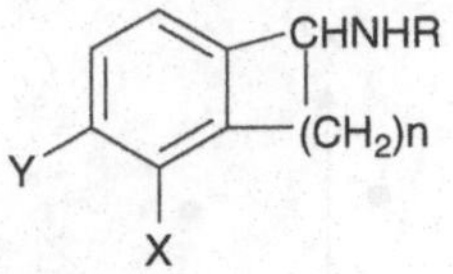

Figure 7-6. General structure of the aminotetralins and aminoindans considered in the discriminant analysis studied by Martin et al. [1974].

Several descriptors were investigated, with two being shown to be statistically significant. These were a Taft steric parameter and an indicator variable that was set to one for substituents at the X position (see Figure 7-6) and zero for substituents at the Y position. Just one molecule was misclassified in the resulting model, which was also confirmed using randomly selected training and test subsets of the data.

3.3   Neural Networks

Neural networks evolved from research into computational models of the brain. The two most commonly used neural network architectures in chemistry are the *feed-forward network* and the *Kohonen network* [Schneider and Wrede 1998; Zupan and Gasteiger 1999]. A feed-forward neural network consists of layers of nodes with connections between all pairs of nodes in adjacent layers, as illustrated in Figure 7-7. Each node exists in a state between zero and one, the state of each node depending on the states of the nodes to which it is connected in the previous layer and the strengths (or weights) of these connections. A key feature of the feed-forward neural network is the presence of one or more layers of hidden nodes; it was the introduction of such multilayer networks together with the back-propagation algorithm used to train them [Rummelhart et al. 1986] that enabled neural networks to be successfully applied to problems in many varied fields.

The neural network must first be "trained"; this is achieved by repeatedly providing it with a set of inputs and the corresponding outputs from a training set. Each node in the input layer may for example correspond to one of the descriptors that are used to characterise each molecule. The weights and other parameters in the neural network are initially set to random values and so the initial output from the neural network (from the output node or nodes) may be significantly different to the desired outputs. A key part of the training phase therefore involves modification of the various weights and parameters that govern the status of the nodes and the connections between them. Once the neural network has been trained it can then be used to predict the values for new, unseen molecules. A practical example of the use of a neural network to predict "drug-likeness" will be

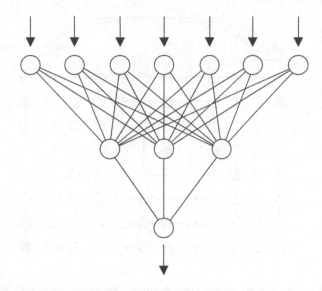

Figure 7-7. A feed-forward neural network with seven input nodes, three hidden nodes and one output node.

discussed in Chapter 8; another common use has been in generating QSAR models [Andrea and Kalayeh 1991; Manallack et al. 1994].

The feed-forward neural network is a supervised learning method, as it uses the values of the dependent variables to derive the model. The Kohonen network or *self-organising map* by contrast is an unsupervised learning method. A Kohonen network most commonly consists of a rectangular array of nodes (Figure 7-8). Each node has an associated vector that corresponds to the input data (i.e. the molecular descriptors). Each of these vectors initially consists of small random values. The data is presented to the network one molecule at a time and the distance between the molecule vector and each of the node vectors is determined using the following distance metric:

$$d = \sum_{i=1}^{p} (x_i - v_i)^2 \qquad (7.5)$$

where v_i is the value of the ith component of the vector **v** for the node in question and x_i is the corresponding value for the input vector. The sum is over the p molecular descriptors. The node that has the minimum distance to the input vector

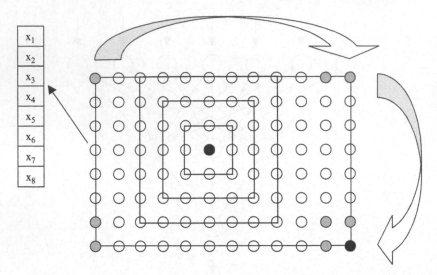

Figure 7-8. Kohonen network showing neighbourhood behaviour. Each node has an associated vector of values. During training the size of the neighbourhood around each winning node is gradually reduced. In order to eliminate edge effects a Kohonen network may be constructed so that opposite sides of the rectangle are joined together, as indicated by the arrows. Thus the immediate neighbours of the bottom right node would be those shaded in grey.

can thus be identified. The vector of this winning node is then updated according to the following formula:

$$v_i' = v_i + \eta (x_i - v_i) \tag{7.6}$$

where η is a gain term. A key feature of the Kohonen network is that in addition to updating the winning node the vectors of its neighbouring nodes are also updated using the same formula. This updating mechanism thus makes the vector of the winning node, and those of its neighbours, more similar to the input. During the training process the gain term gradually decreases, so slowing the rate at which the vectors are modified. The neighbourhood radius is also reduced during training, so localising the activity, as indicated in Figure 7-8.

By modifying the vectors of not only the winning node but also its neighbours the Kohonen network creates regions containing similar nodes. For example, Figure 7-9 shows a Kohonen classification of drug and non-drug molecules [Schneider 2000] (see also Chapter 8). To use a Kohonen network the vector for an unknown molecule would be presented and its grouping identified according to the location of the nearest node. Kohonen networks also provide a mechanism for data reduction and so are an alternative to non-linear mapping.

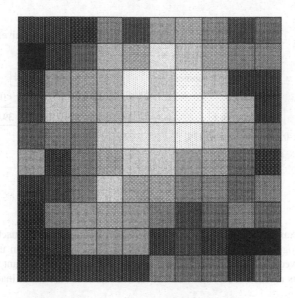

Figure 7-9. Classification of drug and non-drug molecules using a Kohonen network. The distribution of the two types of molecules is indicated by grey shading, with dark regions being dominated by drugs and light regions by non-drugs. (Adapted from Schneider 2000.)

One practical issue with neural networks is the problem of *overtraining*. An overtrained neural network will often give excellent results on the test data but will have poor predictive ability on an unseen data set. This arises because the network "memorises" the training data. In order to try and alleviate this problem the training data can be divided into two; one portion to be used for training as usual, and the other to evaluate the performance of the network during the training phase. It is often observed that the performance of the network in predicting this second evaluation set will initially increase but will then either reach a plateau or start to decline. At this point the network has maximum predictive ability and training should cease.

3.4 Decision Trees

A drawback with feed-forward neural networks is that it is generally not possible to determine why the network gives a particular result for a given input. This is due to the complex nature of the interconnections between the nodes; one cannot easily determine which properties are important. *Decision trees* by contrast are very interpretable as they consist of a set of "rules" that provide the means to associate specific molecular features and/or descriptor values with the activity or property of interest. A decision tree is commonly depicted as a tree-like structure, with each node corresponding to a specific rule. Each rule may

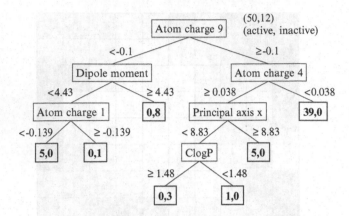

Figure 7-10. Example of a decision tree for a set of 50 active and 12 inactive sumazole and isomazole analogues [A-Razzak and Glen 1992]. In order to classify an unknown molecule a path is followed through the tree according to the values of the relevant properties, until a terminal node is reached. The values indicate the number of active and inactive molecules at each node.

correspond to the presence or absence of a particular feature or to the value of a descriptor. An example of a decision tree is shown in Figure 7-10; this was one of several produced during an investigation into the application of an algorithm called ID3 (see below) to a series of literature QSAR data sets [A-Razzak and Glen 1992]. The decision tree shown in Figure 7-10 is for a series of sumazole and isomazole compounds, some of which possess inotropic properties relevant to cardiac failure. This data set is of particular interest because the active compounds are surrounded in the descriptor space by the inactive molecules; such a situation can cause problems for regression-based methods. To use a decision tree one starts at the root node and follows the edge appropriate to the first rule. This continues until a terminal node is reached, at which point one can assign the molecule into the appropriate activity class. One can also use the decision tree to determine what features give rise to activity and what leads to inactivity.

Various methods are available to construct decision trees. Most of these follow the same basic approach, which is to start with the entire data set and identify the descriptor or variable that gives the "best" split. This enables the data set to be divided into two or more subsets. The same procedure is then applied to each of these subsets, and so on until all of the molecules have been appropriately divided into their distinct classes or until no more splits of significance can be found. One group of algorithms (including ID3, C4.5 and C5.0 [Quinlan 1993; C5.0]) uses results from the field of *information theory* to decide which criteria to choose at each step. If there are N molecules divided into C_1, C_2, C_3 ... classes, each class

C_i containing n_i molecules, then the information corresponding to this distribution (also called the *entropy* of the system) is given by:

$$I = \sum p_i \ln p_i \qquad (7.7)$$

where $p_i = n_i/N$. This result is used to decide which variable or descriptor to use in order to construct the next rule: that which gives rise to the largest increase in entropy (also called the *gain*). In *recursive partitioning* (RP), another popular method for constructing decision trees, the usual node splitting strategy is to split on the descriptor that is statistically the most significant. For example, in a study to investigate the use of RP to explain Monoamine Oxidase activity [Hawkins et al. 1997] in which the compounds were characterised by a 153-element substructural key the most significant descriptor was found to be the presence in the molecule of a carbon–carbon triple bond. A recent study investigated the use of RP in combinatorial library design and HTS analysis, and involved a comparison of various implementations of the algorithm and different descriptor sets [van Rhee et al. 2001]. In this particular case it was concluded that models built from the screening data of a subset of the entire library were able to identify a significant proportion of the active molecules in the remaining library, thereby reducing the number of compounds that needed to be screened subsequently. Two of the more widely used RP algorithms are CART (Classification and Regression Trees [Breiman et al. 1984; CART]) and FIRM (Formal Inference-based Recursive Modeling [Hawkins 1997; FIRM]).

Some decision tree methods only consider binary splits at each node; others can consider multiple branches. Some methods require the descriptors to be either binary (e.g. indicating the presence or absence of a substructure) or to be divided prior to the calculation into distinct categories (e.g. ClogP < 2.0; ClogP ≥ 2.0). Other implementations are able to use both binary and real-valued data and can automatically determine where to make the splits, though at the expense of longer compute times. A problem with some of the basic algorithms is that the resulting tree may be very complex, giving an excellent fit to the training data but whose performance on unseen, test data might be less effective. Some programs are able to prune these complex trees, for example by removing subtrees that have a high estimated prediction error and replacing them by a single node. Single decision trees can also be prone to one particular split assuming undue importance. In order to address these problems a variety of ensemble approaches have been developed; these involve the construction of collections of trees, each of which is generated by training on a subset of the data set. These data subsets may be generated using a bootstrap method in which the data set is sampled with replacement (i.e. there may be duplicates in the subset and some molecules may not appear). The simplest ensemble method is termed *bagging*, in which trees are repeatedly generated on bootstrap samples and new molecules are classified using a majority voting mechanism. *Random forests* [Breiman 2001] are an extension of bagging, in which a small subset of the descriptors is randomly selected at each node rather than using the full set. In *boosting* [Quinlan 1996] each tree is designed to improve the performance for data points misclassified by its predecessor. This is achieved by giving more weight to such misclassified points. The set of trees is then used to classify unseen data using a voting scheme which gives a higher weight to the predictions for these later, more "expert" trees.

The general-purpose decision tree algorithms are able to deal with relatively large data sets but for certain problems more specialised algorithms are required. One such example is a RP procedure called SCAM [Rusinko et al. 1999]. This method uses mathematical and computing techniques that take advantage of the characteristics of binary descriptors. As a consequence it is able to analyse data sets containing hundreds of thousands of molecules, each of which is characterised by very large binary fingerprints derived from procedures such as atom pairs or topological torsions.

The algorithms described above are supervised in nature, with the activity data being used to guide construction of the tree. Similar hierarchical structures can also be constructed using unsupervised methods, as in the *phylogenetic-like tree* method [Nicolaou et al. 2002; Tamura et al. 2002; Bacha et al. 2002]. Central to this particular approach is the use of a MCS algorithm (see Chapter 5). At each iteration of the procedure, cluster analysis is applied to the structures in each node, following which the MCS of the structures in each cluster is determined. Various rules are applied to decide whether this represents a new feature, for example to check that the substructure has not been discovered previously or is a subset of the parent node. If these tests are successful then a new node is added to the tree. This procedure is initially applied to just the active molecules, following which the inactive compounds are filtered through the tree. This enables the average activity of each node to be determined, thereby identifying nodes that may be of particular interest for further visual inspection.

3.5   Support Vector Machines and Kernel Methods

The *support vector machine* (SVM) has become a popular classification technique. The SVM attempts to find a boundary or hyperplane that separates two classes of compounds. The hyperplane is positioned using examples in the training set which are known as the *support vectors*; the use of a subset of the training data prevents overtraining. When the data cannot be separated linearly, *kernel functions* can be used to transform it to higher dimensions where it does become linearly separable. Molecules in the test set are mapped to the same feature space and their activity is predicted according to which side of the hyperplane they fall. The distance to the boundary can be used to assign a confidence level to the prediction such that the greater the distance the higher is the confidence in the prediction. SVMs have been applied to the prediction of activity against a series of G-protein coupled receptors (GPCRs) [Saeh et al. 2005]. They have also been applied in an active learning scenario where compounds are selected for successive rounds of screening [Warmuth et al. 2003]. A disadvantage of the SVM is that, as for neural networks, it is a black box method, making it more difficult to interpret the results than for some methods such as decision trees.

Binary kernel discrimination (BKD) is another machine learning method that uses a kernel function that has been developed for binary data such as 2D fingerprints. BKD was first applied in chemoinformatics by Harper et al. [2001] using the kernel function described below:

$$K_\lambda(i, j) = [\lambda^{n-d_{ij}} (1 - \lambda)d_{ij}]^{\beta/n} \qquad (7.8)$$

where λ is a smoothing parameter, n is the length of the binary fingerprint, d_{ij} is the Hamming distance between the fingerprints for molecules i and j, and $\beta(\beta \leq n)$ is a user-defined constant. The optimum value for λ is determined using a training set and the following equation:

$$S_{\text{BKD}}(j) = \frac{\sum\limits_{i \in \text{actives}} K_\lambda(i, j)}{\sum\limits_{i \in \text{inactives}} K_\lambda(i, j)} \qquad (7.9)$$

A score is computed for each training set molecule based on the other training set molecules for λ values ranging from 0.50 to 0.99. The optimum value of λ is that which gives the minimum sum of ranks of the active molecules. The equation can then be used to rank compounds in a test set. The value of λ effectively determines the number of near neighbours that are used to predict the activity of compounds in the test set. BKD has been applied to the NCI AIDS data set and to agrochemical HTS data sets where it has been shown to be superior to SSA [Wilton et al. 2003]. Since Harper's work the complement of the Tanimoto coefficient (i.e. the Soergel distance) has been shown to outperform the Hamming distance over a range of different data sets [Chen et al. 2006].

HTS data is often rather "noisy" and it is therefore of interest to determine how robust are the various machine learning methods. For example, Chen et al. [2006] have simulated such data by progressively reassigning known actives as inactives, and vice versa. This effectively introduces increasing numbers of false positives and false negatives. BKD was compared with SSA and although BKD was found to be more effective when presented with the clean data, it was much less effective than SSA for the noisy data. SSA was found to be robust to noisy data even when up to 90% of the actives consisted of false positives. A similar study compared a NBC with SVM and RP [Glick et al. 2006]. Comparable results were found with the SVM being most effective with clean data, and the NBC (which is similar to SSA) being more robust when presented with noisy data.

4.   SUMMARY

The development of methods for the analysis of HTS data is an area of much interest and current research. Some of the computational tools that are being used to tackle this problem are well established; however, other techniques that have been applied in the wider data mining community are of growing interest such as SVMs and other kernel-based learning methods [Christianini and Shawe-Taylor 2000; Bennett and Campbell 2000; Harper et al. 2001]. Of all the areas discussed in this book, the analysis of large diverse data sets is probably the least advanced with most scope for further developments. This is of course largely due to the fact that the experimental techniques are relatively new. Moreover, advances in the experimental techniques themselves can also have a major impact on the approach that one takes to the subsequent data analysis, such as the use of statistical methods to assess the underlying quality of the data [Zhang et al. 1999]. It will be particularly important to find ways to incorporate such information into any subsequent chemoinformatics analysis. Indeed, it is likely that no single analysis technique will prove to be universally recommended, but that it will be necessary to consider a number of different methods to determine which is most appropriate, or alternatively to use several techniques in parallel. An example of the latter approach is the published analysis of the NCI's anti-cancer data [Shi et al. 2000]. This data set comprises approximately 70,000 compounds which have been tested in 60 human cancer cell lines. A variety of methods were employed, including principal components analysis, cluster analysis, multidimensional scaling and neural networks. One finding of particular interest was that although the data for any single cell line were not particularly informative, patterns of activity across the 60 cell lines did provide conclusive information on the mechanisms of action for the screened compounds.

Chapter 8

VIRTUAL SCREENING

1. INTRODUCTION

Virtual screening is the computational or *in silico* analogue of biological screening. The aim of virtual screening is to score, rank and/or filter a set of structures using one or more computational procedures. Virtual screening is used, for example, to help decide which compounds to screen, which libraries to synthesise and which compounds to purchase from an external company. It may also be employed when analysing the results of an experiment, such as a HTS run.

There are many different criteria by which the structures may be scored, filtered, or otherwise assessed in a virtual screening experiment. For example, one may use a previously derived mathematical model such as a multiple linear regression equation to predict the biological activity of each structure. One may use a series of substructure queries to eliminate molecules that contain certain undesirable functionality. If the structure of the target protein is known, one may use a docking program to identify structures that are predicted to bind strongly to the protein active site. The number of structures that may need to be considered in virtual screening experiments can be very large; when evaluating virtual combinatorial libraries the numbers may run into the billions. It is therefore important that the computational techniques used for the virtual screening have the necessary throughput. This is why it may often be most effective to use a succession of virtual screening methods of increasing complexity [Charifson and Walters 2000]. Each method acts as a filter to remove structures of no further interest, until at the end of the process a series of candidate structures are available for final selection, synthesis or purchase.

Wilton et al. [2003] have suggested that there are four main classes of virtual screening methods, according to the amount of structural and bioactivity data available. If just a single active molecule is known then similarity searching can be performed in what is sometimes known as ligand-based virtual screening. If several actives are available then it may be possible to identify a common 3D pharmacophore, followed by a 3D database search. If a reasonable number of active and inactive structures are known they can be used to train a machine learning technique such as a neural network which can then be used for virtual

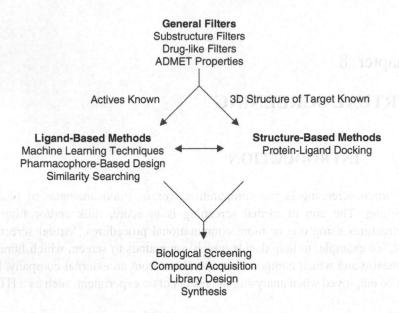

Figure 8-1. Many virtual screening processes involve a sequence of methodologies.

screening. Finally, protein–ligand docking can be employed when the 3D structure of the protein is known. In addition to these methods that are specific to particular targets, virtual screening techniques have been developed that are of more general applicability. These include, for example, methods that attempt to predict the likelihood that a molecule possesses "drug-like" characteristics and that it has the desired physiochemical properties to be able to reach its site of action within the body. A schematic illustration of a typical virtual screening flowchart is shown in Figure 8-1.

Similarity searching and pharmacophore methods were described in Chapters 5 and 2, respectively. Many of the machine learning techniques that can be applied when several actives and inactives are known were introduced in Chapter 7 and will be further expanded here in the context of predicting "drug-likeness". We will then discuss the application of protein–ligand docking in structure-based virtual screening, and finally the prediction of ADMET properties.

2. "DRUG-LIKENESS" AND COMPOUND FILTERS

The advent of combinatorial chemistry and HTS enabled much larger number of compounds to be synthesised and tested but it did not lead to the expected improvements in the numbers of lead molecules being identified. This observation sparked much interest in the concept of "drug-likeness" [Clark and Pickett 2000;

Walters and Murcko 2002] and attempts to determine which features of drug molecules confer their biological activity and distinguish them from general "organic" compounds.

Among the simplest types of methods that can be used to assess "drug-likeness" are substructure filters. As discussed in Chapter 7, a compound collection may include molecules that contain reactive groups known to interact in a non-specific manner with biological targets, molecules that give "false positive" results due to interference with certain types of biological assay, or molecules which are simply inappropriate starting points for a drug discovery programme [Roche et al. 2002]. Many of these features can be defined as substructures or as substructure counts that are used to filter both real and virtual data sets. One may wish to apply such filters to the output from a HTS run, in order to eliminate known problem molecules from further consideration (of course, one would prefer to eliminate such compounds prior to the screen). They are also extremely useful when designing virtual libraries and when selecting compounds to purchase from external suppliers. One should, however, always remember that such filters tend to be rather general in nature and that for any specific target it may be necessary to modify or extend them accordingly.

Other approaches to the question of "drug-likeness" were derived by analysing the values of relatively simple properties such as molecular weight, the number of rotatable bonds and the calculated $\log P$ in known drug molecules. Considerations such as these led to the formulation of the "rule of five" [Lipinski et al. 1997] which constitutes a set of simple filters that suggest whether or not a molecule is likely to be poorly absorbed. The "rule of five" states that poor absorption or permeation is more likely when:

1. The molecular weight is greater than 500
2. The $\log P$ is greater than five
3. There are more than five hydrogen bond donors (defined as the sum of OH and NH groups)
4. There are more than ten hydrogen bond acceptors (defined as the number of N and O atoms)

Excluded from this definition are those compounds that are substrates for biological transporters. An obvious attraction of this model is that it is extremely simple to implement and very fast to compute; many implementations report the number of rules that are violated, flagging or rejecting molecules that fail two or more of the criteria. A more extensive evaluation of property distributions for a set of drugs and non-drugs has identified the most likely values for "drug-like" molecules [Oprea 2000]. For example, 70% of the "drug-like" compounds had between zero and two hydrogen bond donors, between two and nine hydrogen bond acceptors, between two and eight rotatable bonds and between one and four rings. The "rule of five" was derived following a statistical analysis of known drugs; a similar analysis has since been carried out on agrochemicals with modified sets of rules being derived that relate to the properties of herbicides and insecticides [Tice 2001].

Others have attempted to derive more sophisticated computational models of "drug-likeness" using techniques such as neural networks or decision trees. Typically these models start with a training set of drugs and non-drugs, for which a variety of descriptors are calculated. The training set and its corresponding descriptors are then used to develop the model, which is evaluated using a test set. For example, Sadowski and Kubinyi constructed a feed-forward neural network with 92 input nodes, 5 hidden nodes and 1 output node to predict "drug-likeness" [Sadowski and Kubinyi 1998]. The data set comprised compounds from the WDI (the drugs) and a set of structures extracted from the Available Chemicals Directory [ACD] (assumed to have no biological activity, and therefore to be non-drugs). Each molecule was characterised using a set of atom types originally devised by Ghose and Crippen for the purposes of predicting $\log P$ [Ghose and Crippen 1986]. The counts of each of the 92 atom types for the molecules provided the input for the neural network. These descriptors act as a form of extended molecular formula and were found to perform better than whole-molecule descriptors such as the $\log P$ itself or detailed descriptors such as a structural key or hashed fingerprint. The network was able to correctly assign 83% of the molecules from the ACD to the non-drugs class and 77% of the WDI molecules to the drugs class. Other groups have obtained comparable results [Ajay et al. 1998; Frimurer et al. 2000].

Wagener and van Geerestein [2000] used decision trees to tackle this problem. The same databases were used to identify drug and non-drug molecules and the same set of Ghose-Crippen atom type counts were used to characterise each molecule. The C5.0 algorithm [Quinlan 1993; C5.0] was employed. Its performance was comparable to that of the neural network, correctly classifying 82.6% of an independent validation set. A second model designed to reduce the false negative rate (i.e. the misclassification of drug molecules) was able to correctly classify 91.9% of the drugs but at the expense of an increased false positive rate (34.3% of non-drugs misclassified). Some of the rules in the decision tree were of particular interest; these suggested that merely testing for the presence of some simple functional groups such as hydroxyl, tertiary or secondary amino, carboxyl, phenol or enol groups would distinguish a large proportion of the drug molecules. Non-drug molecules were characterised by their aromatic nature and a low functional group count (apart from halogen atoms).

Gillet et al. [1998] used a genetic algorithm to build a scoring scheme for "drug-likeness". The scoring scheme was based on the following physicochemical properties: molecular weight; a shape index (the kappa–alpha 2 index [Hall and Kier 1991]) and numbers of the following substructural features: hydrogen bond donors; hydrogen bond acceptors; rotatable bonds; and aromatic rings. Each property was divided into a series of bins that represent ranges of values of the property, such as molecular weight ranges or counts of the number of times a particular substructure or feature occurs in a molecule. A weight is associated with each bin and a molecule is scored by determining its property values and then summing the appropriate weights across the different properties. The genetic algorithm was used to identify an optimum set of weights such that maximum discrimination between the two classes was achieved, with molecules in one class scoring highly while molecules in the other class have low scores. In the case of "drug-likeness" the genetic algorithm was trained using a sample of the SPRESI database [SPRESI] to represent "non-drug-like" compounds and a sample of the WDI to represent "drug-like" compounds. The resulting model was surprisingly effective at distinguishing between the two classes of compounds, as can be seen in Figure 8-2. It has subsequently been used to filter compounds prior to high-throughput screening [Hann et al. 1999].

One interesting development has been the introduction of "lead-likeness" as a concept distinct from "drug-likeness". The underlying premise is that during the optimisation phase of a lead molecule to give the final drug there is an increase in the molecular "complexity", as measured by properties such as molecular weight, the numbers of hydrogen bond donors and acceptors and ClogP. It has therefore been argued [Teague et al. 1999; Hann et al. 2001] that one should

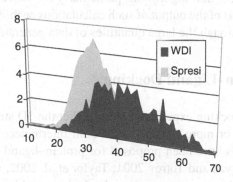

Figure 8-2. Output from the genetic algorithm scoring scheme used to distinguish drugs from non-drugs showing the degree of discrimination that can be achieved. The WDI contains known drug-like molecules whereas SPRESI is a database containing general "organic" molecules, assumed to have no biological activity.

use "lead-like" criteria when performing virtual screening at that stage rather than the "drug-like" criteria typified by the "rule of five". These arguments are supported by analyses of case-histories of drug discovery together with theoretical models of molecular complexity. Interest in lead-likeness led in turn to fragment-based approaches to drug discovery, wherein less complex molecules are screened to provide starting points for subsequent optimisation. The small size of the molecules used in such approaches means that they need to be screened at higher concentrations or using biophysical techniques such as x-ray crystallography or NMR. In addition to the practical aspects of fragment-based drug discovery there have also been associated theoretical developments, such as the "rule of three" [Congreve et al. 2003] (a fragment equivalent of the rule of five) and the concept of ligand efficiency [Hopkins et al. 2004] (a method for prioritising the output from screening experiments in order to identify the most promising initial candidates).

3.   STRUCTURE-BASED VIRTUAL SCREENING

As the number of protein crystal structures has increased so too has the interest in using this detailed structural knowledge for library design, compound acquisition and data analysis. Structure-based design methods have been developed over many years. However, much of this activity was geared towards methods for the detailed analysis of small numbers of molecules. Several factors have contributed to the move towards higher-throughput structure-based methods. First, high-performance computer hardware (often based on the linux operating system) has provided unparalleled amounts of dedicated computing power at a low cost to researchers. Second a significant effort has been expended in the development of new algorithms, particularly for molecular docking. Finally, tools for the analysis of the output of such calculations enable scientists to navigate more effectively through the large quantities of data generated.

3.1   Protein–Ligand Docking

The aim of a docking experiment is to predict the 3D structure (or structures) formed when one or more molecules form an intermolecular complex. A large number of methods have been proposed for protein–ligand docking [Blaney and Dixon 1993; Abagyan and Totrov 2001; Taylor et al. 2002; Halperin et al. 2002]. There are essentially two components to the docking problem. First, it is necessary to have a mechanism for exploring the space of possible protein–ligand geometries (sometimes called *poses*). Second it is necessary to be able to score or rank these poses in order to identify the most likely binding mode for each compound and to assign a priority order to the molecules.

The difficulty with protein–ligand docking is in part due to the fact that it involves many degrees of freedom. The translation and rotation of one molecule relative to another involves six degrees of freedom. There are in addition the conformational degrees of freedom of both the ligand and the protein. The solvent may also play a significant role in determining the protein–ligand geometry and the free energy of binding even though it is often ignored. An expert computational chemist may be able to predict the binding mode of a ligand using interactive molecular graphics if he/she has a good idea of the likely binding mode (e.g. if the x-ray structure of a close analogue is available). However, manual docking can be very difficult when dealing with novel ligand structures and is clearly impractical for large numbers of molecules.

It can be convenient to classify the various docking algorithms according to the degrees of freedom that they consider. Thus the earliest algorithms only considered the translational and rotational degrees of freedom of protein and ligand, treating each as rigid bodies. The algorithms most widely used at present enable the ligand to fully explore its conformational degrees of freedom; some programs do also permit some very limited conformational flexibility to the protein (e.g. permitting the side chain protons of serine, threonine, trypsin and lysine residues to freely rotate). A few published algorithms attempt to take more of the protein flexibility into account, but these have not yet been extensively evaluated nor are they in widespread use [Carlson 2002; Shoichet et al. 2002].

The DOCK algorithm developed by Kuntz and co-workers is generally considered one of the major advances in protein–ligand docking [Kuntz et al. 1982, 1994; Desjarlais et al. 1988; Kuntz 1992]. The earliest version of the DOCK algorithm only considered rigid body docking and was designed to identify molecules with a high degree of shape complementarity to the protein binding site. The first stage of the DOCK method involves the construction of a "negative image" of the binding site. This negative image consists of a series of overlapping spheres of varying radii, derived from the molecular surface of the protein. Each sphere touches the molecular surface at just two points (see Figure 8-3). Ligand atoms are then matched to the sphere centres so that the distances between the atoms equal the distances between the corresponding sphere centres, within some tolerance. These pairs of ligand atoms and sphere centres can be used to derive a translation–rotation matrix that enables the ligand conformation to be oriented within the binding site using molecular fitting. The orientation is checked to ensure that there are no unacceptable steric interactions and it is then scored. New orientations are produced by generating new sets of matching ligand atoms and sphere centres. The procedure continues until all possible matches have been considered.

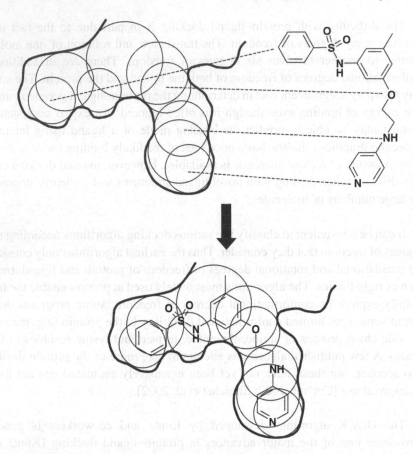

Figure 8-3. Operation of the DOCK algorithm. A set of overlapping spheres is used to create a "negative image" of the active site. Ligand atoms are matched to the sphere centres, thereby enabling the ligand conformation to be oriented within the active site for subsequent scoring.

More recent algorithms that take the ligand conformational degrees of freedom into account can be conveniently classified according to the way in which they explore the conformational space. The simplest way is to generate an ensemble of ligand conformations using an appropriate conformational search algorithm (see Chapter 2) and then to dock each of these using a rigid body algorithm. This may seem to be a rather inefficient approach, but nevertheless some quite effective methods are based on this method; these usually employ a variety of filters and approximations to rapidly identify the conformations of interest [Glide, FRED]. The relevance of a conformational analysis performed using the isolated ligand might be questioned but a number of studies have demonstrated that the conformations of ligands observed in protein–ligand complexes often have very similar geometries to minimum-energy conformations of the isolated ligand [Boström et al. 1998]. Other methods explore the orientational and conformational degrees of freedom at the same time. Most of these methods fall into one of three categories: Monte Carlo algorithms, genetic algorithms and incremental construction approaches.

The simplest Monte Carlo algorithms for protein–ligand docking are closely related to those employed for conformational analysis [Goodsell and Olson 1990]. At each iteration of the procedure either the internal conformation of the ligand is changed (by rotating about a bond) or the entire molecule is subjected to a translation or a rotation within the binding site. The new configuration is accepted if its energy (V_{new}) is lower than that of its predecessor (V_{old}) or if the Boltzmann factor $\exp[-(V_{new} - V_{old})/kT]$ is greater than a random number between zero and one. Simulated annealing (also described in Chapter 2) is frequently employed to search for the lowest energy solution. More complex Monte Carlo methods such as *tabu search* have also been used; this particular method directs the search away from regions already visited, thereby ensuring greater exploration of the binding site [Baxter et al. 1998].

Genetic and evolutionary algorithms can be used to perform protein–ligand docking [Judson et al. 1994; Jones et al. 1995b; Oshiro et al. 1995; Gehlhaar et al. 1995]. In these methods each chromosome in a population encodes one conformation of the ligand together with its orientation within the binding site. A scoring function is used to calculate the fitness of each member of the population and to select individuals for each iteration. As with the Monte Carlo search methods, the underlying random nature of the genetic algorithm means that it is usual to perform a number of runs and to select the structures with the highest scores.

Incremental construction methods construct conformations of the ligand within the binding site in a series of stages [Leach and Kuntz 1990; Welch et al. 1996; Rarey et al. 1996]. A typical algorithm of this type first identifies one or more "base fragments" which are docked into the binding site. These base fragments are usually fairly sizeable (often rigid) parts of the molecule such as ring systems. The orientations of the base fragment then form the basis for a systematic conformational analysis of the remainder of the ligand, with the protein binding site providing an additional set of constraints that can be used to prune the search tree.

3.2   Scoring Functions for Protein–Ligand Docking

It is often useful to make a distinction between docking and scoring in structure-based virtual screening experiments. Docking involves the prediction of the binding mode of individual molecules, the aim being to identify the orientation that is closest in geometry to the observed (x-ray) structure. Several studies have been performed to evaluate the performance of docking programs using data sets derived from the PDB and it is found that these programs are able to correctly predict the binding geometry in more than 70% of the cases [Jones et al. 1997; Kramer et al. 1999; Nissink et al. 2002]. However, it is not always straightforward to predict which of the many docking programs will give the best result in any particular case [Kontoyianni et al. 2004; Warren et al. 2006]. An illustration of the type of results that are produced by a typical docking program is shown in Figure 8-4. For virtual screening, however, it is also necessary to be able to score or rank the ligands using some function related to the free energy of association of the protein and ligand to form the intermolecular complex. Ideally, of course, the same function would be used for both docking the ligands and for predicting their free energies of binding. This is indeed the case for some methods. Others use different functions for docking and for scoring. This may be due to the fact that the large number of orientations generated during a typical docking run requires a function that can be calculated very rapidly; the time required to calculate some scoring functions precludes their use during the docking phase. They are instead used to score the final structures produced by the docking algorithm. The use

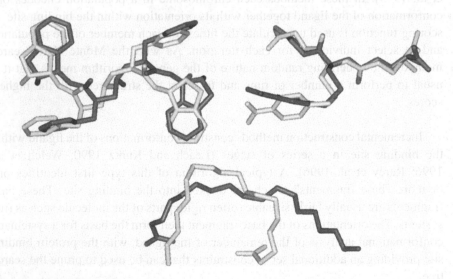

Figure 8-4. Illustration of the range of results produced by a typical docking program. Here we show the results obtained by running the GOLD program [Jones et al. 1995b] on three ligands from the PDB. In each case the x-ray conformation is shown in dark grey and the top-ranked docking result in light grey. The PDB codes for the three ligands are (clockwise, from top left) 4PHV (a peptide-like ligand in HIV Protease), 1GLQ (a nitrophenyl-substituted peptide in glutathione S transferease) and 1CIN (oleate in fatty acid binding protein). These dockings were classified as "good", "close" and "wrong" [Jones et al. 1997].

of separate functions for docking and for scoring is also a reflection of the fact that many of the functions that perform well in predicting the binding mode of individual molecules do not necessarily perform well for scoring and ranking different ligands binding to the same protein. Moreover, it is generally found that whilst docking and scoring methods are often able to achieve some degree of enrichment (compared to random) they are rarely able to accuractely predict the free energies of binding [Leach et al. 2006].

Most of the scoring functions in current use are consistent with the hypothesis that the free energy of binding can be written as a linear summation of terms to reflect the various contributions to binding [Böhm and Klebe 1996]. A complete equation of this type would have the following form [Ajay and Murcko 1995]:

$$\Delta G_{bind} = \Delta G_{solvent} + \Delta G_{conf} + \Delta G_{int} + \Delta G_{rot} + \Delta G_{t/r} + \Delta G_{vib} \qquad (8.1)$$

$\Delta G_{solvent}$ captures contributions to the free energy of binding from solvent effects. These contributions arise from interactions between the solvent and the ligand, protein and protein–ligand complex. Most approaches to docking and scoring are performed in the absence of any explicit solvent molecules; it is thus necessary to devise ways to take these important effects into account. ΔG_{conf} arises from conformational changes to the protein and to the ligand. This is difficult to estimate for the protein. The ligand moves from an ensemble of conformations in free solvent to what is often assumed to be a single dominant conformation in the intermolecular complex (that may not necessarily be a minimum-energy conformation for the isolated ligand). The penalty associated with this can be estimated in various ways; one study suggested an average penalty of 3 kcal/mol relative to the most significant conformation in solution [Boström et al. 1998]. ΔG_{int} is the contribution from protein–ligand interactions, arising from electrostatic and van der Waals forces. ΔG_{rot} is the penalty associated with freezing internal rotations of the protein and the ligand. It is largely entropic in nature; a very simple estimate of its value is to assume that there are three isoenergetic states for each rotatable bond, leading to a loss of $RT\ln 3$ (~ 0.7 kcal/mol) for each frozen rotatable bond. $\Delta G_{t/r}$ is the loss in translational and rotational degrees of freedom arising from the association of two bodies to give a single body. It is often assumed to be constant for all ligands and so is generally ignored when calculating relative binding energies for different ligands to the same protein. ΔG_{vib} is the free energy due to changes in vibrational modes. It is hard to calculate and usually ignored.

Some very simple functions have been used for scoring. Two examples are the DOCK function [Desjarlais et al. 1988] and the piecewise linear potential [Gelhaar et al. 1995] whose functional forms are illustrated in Figure 8-5. The

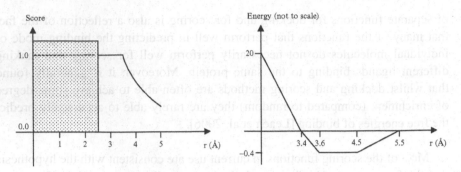

Figure 8-5. Two simple scoring functions used in docking. On the left is the basic scoring scheme used by the DOCK program [Desjarlais et al. 1988]. On the right is the piecewise linear potential with the parameters shown being those used to calculate steric interactions [Gelhaar et al. 1995]. (Adapted from Leach 2001.)

DOCK function scores 1.0 for atom pairs separated by 2.3–3.5 Å. Between 3.5 and 5.0 Å the score decays in an exponential fashion. Atom pairs separated by more than 5.0 Å score zero. The piecewise linear potential divides the interaction into attractive and repulsive regions with different sets of parameters for steric and hydrogen bonding interactions. Despite the simple nature of these functions they often perform as well as much more complex functions in comparative studies. Force fields (see Appendix 2) can also be used for docking and scoring; however, these only provide some of the contributions listed in Equation 8.1. Thus they may perform well for certain systems but are not generally considered to be widely applicable. One potential drawback of both the simple scoring functions and force field scoring is that the function is typically evaluated for all pairs of ligand and protein atoms that are within some cut-off distance. This calculation can be speeded up by pre-calculating the potentials due to the protein on a regular grid that covers the binding site [Meng et al. 1992]. This makes the calculation of the function proportional to the number of ligand atoms, rather than being proportional to the product of the number of ligand atoms and the number of protein atoms.

A key advance in the development of scoring functions for high-throughput docking and virtual screening applications was the introduction by Böhm of a linear function in which the various terms could be calculated rapidly [Böhm 1994]. The original formulation of this function is as follows:

$$\Delta G_{bind} = \Delta G_0 + \Delta G_{hb} \sum_{h-bonds} f(\Delta R, \Delta \alpha)$$

$$+ \Delta G_{ionic} \sum_{\substack{ionic \\ interactions}} f(\Delta R, \Delta \alpha) \qquad (8.2)$$

$$+ \Delta G_{lipo} |A_{lipo}| + \Delta G_{rot} NROT$$

The function thus includes contributions from hydrogen bonding, ionic interactions, lipophilic interactions and the loss of internal conformational freedom of the ligand. The hydrogen bonding and ionic terms are both dependent on the geometry of the interaction, with large deviations from ideal geometries being penalised. The lipophilic term is proportional to the contact surface area (A_{lipo}) between protein and ligand involving non-polar atoms. The conformational entropy term is directly proportional to the number of rotatable bonds in the ligand (NROT).

The coefficients in Equation 8.2 were determined using multiple linear regression on experimental binding data for 45 protein–ligand complexes. The model reproduced the experimental data with a standard deviation of 1.7 kcal/mol. A number of alternative functions have been derived that are based on this approach; they differ in the experimental data used to derive the model and in the terms included in the linear expansion [Head et al. 1996; Eldridge et al. 1997; Böhm 1998; Murray et al. 1998]. A common modification of Böhm's original formulation is to divide the surface area contributions into three, corresponding to polar/polar, polar/non-polar and non-polar/non-polar interactions.

An alternative strategy to the model-building approach introduced by Böhm is to use a so-called *knowledge-based* method [Muegge and Martin 1999; Mitchell et al. 1999; Gohlke et al. 2000; Sotriffer et al. 2002]. In such an approach, the interaction potential between each ligand–protein atom pair is calculated as a *potential of mean force*. This gives the free energy of interaction of the pair of atoms as a function of their separation. These potentials of mean force are calculated via statistical analyses of the distributions observed in known protein–ligand complexes. An advantage of such an approach is that physical effects that are difficult for other methods to model are automatically taken into account. However, there are some technical issues with the derivation of such potentials, such as the need to define a reference distribution and the treatment of solvation effects [Muegge 2000; Gohlke and Klebe 2001].

There is as yet no universally applicable scoring function able to reproduce experimental binding affinities with an acceptable error for a wide variety of protein targets and ligand structures [Böhm and Stahl 1999; Bissantz et al. 2000; Lyne 2002]. This is perhaps not so surprising when one considers that a change of just 1.4 kcal/mol in the free energy corresponds to a tenfold change in the free energy of binding. Small errors in the scoring function thus correspond to large errors in the dissociation constant. Another important consideration is that the ligands identified using virtual screening are most likely to have rather modest potency, with dissociation constants in the micromolar range. This contrasts with the ligands most commonly used to derive and test current scoring functions; these often bind very tightly to their target protein (with nanomolar dissociation constants). In addition, the ligands currently available in the PDB represent a rather limited chemical diversity, with a large proportion consisting of cofactors, naturally occurring inhibitors, peptides and peptide mimetics. Moreover, a disproportionate number of these complexes are for just a few proteins such as the HIV protease and thrombin. There has been a gradual increase in the rate at which protein–ligand complexes containing "drug-like" inhibitors become available, but it will undoubtedly take some time before a truly comprehensive knowledge base exists. A number of studies have been published assessing the ability of various scoring functions to predict experimentally measured binding affinities [So et al. 2000; Bissantz et al. 2000; Ha et al. 2000; Stahl and Rarey 2001]. One interesting development is the introduction of "consensus scoring" schemes, in which the results from more than one scoring function are combined [Charifson et al. 1999; Terp et al. 2001]. It has been shown that these approaches give better results than just using any of the individual scoring functions.

3.3   Practical Aspects of Structure-Based Virtual Screening

Structure-based virtual screening is a computationally intensive and complex procedure. Docking programs tend to have a multitude of possible parameters that govern their operation; an inappropriate choice may lead to excessively long computation times or even scientifically invalid results. It is usually advisable to spend a reasonable amount of time investigating the various options and performing validation runs on known structures to identify the most appropriate settings.

The ligand structures for structure-based virtual screening usually originate as a series of connection tables or SMILES strings. However, most docking programs require a reasonable 3D conformation as the starting point. Factors that may need to be considered when generating these starting conformations include the geometry of undefined chiral centres, and the ionisation and tautomeric state of the ligand. Some docking and scoring programs require partial atomic charges to be computed for the protein and each ligand; again, the results may be dependent upon the method chosen to perform this calculation.

Preparation of the protein is also important. Hydrogens are not usually visible in protein x-ray structures but most docking programs require protons to be included in the input file. Algorithms are available for automatically predicting the proton positions within the protein itself [Brunger and Karplus 1988; Bass et al. 1992]; some of these can take account of potential hydrogen bonding networks within the protein, but it is usually advisable to inspect the output carefully. The issue of ionisation and tautomeric state is also very relevant to the protein. Having prepared the protein it is necessary to define the binding site. Again, automatic procedures are available for performing this, but these are by no means foolproof. Too small a binding site may mean that some potential ligands will be discarded because they are deemed not to fit; too large and much computational time will be spent exploring unproductive regions of the search space.

Given the computationally intensive nature of docking, the use of appropriate computational filters prior to the docking to eliminate undesirable or inappropriate structures is particularly important. Knowledge about possible binding modes may also be used to enhance the efficiency of the docking procedure. For example, one may wish all solutions to form a key hydrogen bonding or electrostatic interaction. Most docking programs provide mechanisms for introducing such constraints. One may be able to use a 3D pharmacophore search as a filter prior to the docking. As discussed in Chapter 2, it is possible to define 3D pharmacophores containing features such as "excluded regions" or "inclusion volumes" that provide a reasonably accurate representation of the binding site. Alternatively, one may use one of the faster docking programs to reduce the size of the data set prior to a more thorough analysis using a slower but more accurate program.

Following the docking and scoring calculations it is then necessary to analyse the results in order to identify those structures to progress to the next phase. In many respects the analysis of a large-scale docking run is similar to the analysis of a HTS run. The most basic approach would be to simply select the top-scoring ligands for progression. However, it is generally advisable to undertake some form of data analysis, coupled to 3D interactive visualisation of the ligands docked into the binding site. Computational tools for filtering the final docked orientations using relatively simple parameters related to protein–ligand complexes may also be valuable in this final analysis and selection phase. In one such approach [Stahl and Böhm 1998] properties such as the number of close protein–ligand contacts and the fraction of ligand volume buried in the binding pocket were computed. It was observed that the values of these properties for docked structures often deviated significantly from the values for the corresponding protein–ligand x-ray structures. This enabled a series of rules to be developed for the rapid postprocessing of the docking results.

An increasing number of "success stories" describing the use of docking in lead discovery are now appearing in the literature [Lyne 2002]. These demonstrate the widespread adoption of the methodology but also highlight some of the shortcomings, particularly with regards to the scoring functions currently available. An interesting experiment was performed by Doman et al. (2002) who compared HTS and docking. The biological target in this case was the enzyme Tyrosine Phosphatase-1B, a potential drug target for diabetes. A corporate collection containing 400,000 compounds was tested using HTS and molecular docking was used to virtually screen a database of 235,000 commercially available compounds from which 365 were purchased for testing. Nearly 85 of the compounds from the HTS run had an activity less than 100 μM (i.e. 0.021%); 127 of the 365 purchased compounds were active at this level (34.8%). Of particular interest was that the two sets of active compounds were very different from each other, suggesting that the two approaches complemented each other.

Docking and other structure-based design methods will undoubtedly continue to be of much interest as the numbers of crystal structures of proteins of therapeutic interest grows. Although docking is a computationally intensive procedure compared to many other techniques it is very straightforward to implement in a parallel computing environment by the simple expedient of splitting the data set according to the number of processors available. This makes docking particularly relevant for some of the latest computer hardware such as multiprocessor "farms" and Grid-computing. The latter is particularly interesting, often involving the use of spare processing capacity on the personal computers of a corporation's employees or even the entire internet [Davies et al. 2002; Richard 2002].

4.   THE PREDICTION OF ADMET PROPERTIES

It is not sufficient for a drug molecule to bind tightly to its biological target in an *in vitro* assay. It must also be able to reach its site of action *in vivo*. This usually involves passing through one or more physiological barriers such as a cell membrane or the blood-brain barrier. The molecule must remain in the body for an appropriate period of time to give the desired effect but must ultimately be removed from the body by metabolism, excretion or other pathways. Moreover, neither the drug nor its metabolites should be toxic. ADMET (Absorption, Distribution, Metabolism, Elimination and Toxicity) is often used to refer to such aspects of drug discovery; ADMET properties were traditionally considered during the development phase of a drug candidate. There is now increasing interest in taking ADMET properties into account at an earlier stage during drug discovery, in recognition that it may be better (and cheaper) to identify and address any issues earlier in the process rather than later [Hodgson 2001; van de Waterbeemd et al. 2001].

Various ADMET properties have attracted particular attention from the computational chemistry and QSAR communities [Ekins et al. 2000; van de Waterbeemd 2002; Clark and Grootenhuis 2002; Beresford et al. 2002]. The desired route of administration is usually via the oral route and so the assessment of oral bioavailability is important [Veber et al. 2002]. Of the many factors that may enable a substance to be orally bioavailable, permeability through the intestinal cell membrane is often a key requirement [Egan and Lauri 2002]. Another physiological barrier is the blood-brain barrier between the systemic circulation and the brain. The ability to pass through this barrier is a desired property in the case of biological targets in the brain; for other targets the opposite property might be required in order to reduce possible side effects [Norinder and Haeberlein 2002]. In addition to these passive mechanisms of permeability there also exist several transport mechanisms which actively promote (or reduce) the absorption of substances across these barriers; the P-glycoprotein (PGP) efflux pump has been subjected to particular attention [Stouch and Gudmundsson 2002]. Aqueous solubility is a key physicochemical property upon which properties such as oral bioavailability depend [Lipinski 2000; Jorgensen and Duffy 2002]. The cytochrome P450 enzymes play a key role in metabolism and so it is of considerable interest to determine which structures will be substrates or inhibitors of these enzymes [de Groot and Ekins 2002]. Finally, toxicity prediction is a problem of long-standing interest [Greene 2002].

A common problem with many developability properties is the paucity of experimental data, especially human *in vivo* data. This can make it very difficult to build reliable models that can be applied to a wide variety of chemical classes. For this reason there has been significant interest in the identification of *in vitro* systems that are surrogates for the "real" system. An example is the use of Caco-2 cells (derived from a human intestinal cell line) to measure permeability as a surrogate for the human absorption process.

Many of the computational models published to date are "traditional" QSAR models derived using multiple linear regression. However, other methods such as PLS, decision trees, discriminant analysis and neural networks are being increasingly employed. As we have discussed these methods in previous chapters the focus of our subsequent discussion will be on some of the unique descriptors that have been used to construct ADMET models.

4.1 Hydrogen Bonding Descriptors

Several ADMET models are dependent upon the hydrogen bonding capacity of a molecule. In the simplest case this involves simply counting the numbers of donors and acceptors as in Lipinski's "rule of five". More sophisticated measures of hydrogen bonding capacity have also been developed. Abraham has proposed a mechanism for calculating the overall hydrogen bonding acidity or donor propensity (A) and basicity or acceptor propensity (B) of any solute molecule [Abraham et al. 2002]. These parameters were originally obtained by fitting to relevant experimental data such as partition coefficients measured for different water-solvent systems or by analogy to closely related compounds. In addition, a fragment-based approach has been developed with 51 fragment values for A and 81 fragment values for B [Platts et al. 1999]. This is more suited to virtual screening. In the Abraham approach these hydrogen bonding descriptors are then combined with other descriptors using multiple linear regression to derive QSAR equations for a wide variety of physicochemical and ADMET properties [Abraham 1993; Abraham and Platts 2001]. An alternative approach to hydrogen bonding was taken by Jorgensen and colleagues based originally on Monte Carlo simulations of solutes in water. From these simulations a variety of descriptors were derived, including the average number of hydrogen bonds formed over the course of the simulation between the solute and solvent (counting solute donor and solute acceptor hydrogen bonds separately). Note that these hydrogen bond counts can be non-integral. These descriptors were then combined using multiple linear regression to derive models of properties such as solubility, the octanol/water partition coefficient and blood-brain barrier permeability [Duffy and Jorgensen 2000a, b]. A drawback of this simulation-based approach is that it is slow, and so algorithms were devised to estimate these descriptors rapidly, particularly the hydrogen bond counts. These algorithms take into account the electronic and steric environment of possible hydrogen bonding sites and the possibility of intramolecular hydrogen bonding [Jorgensen and Duffy 2002].

4.2   Polar Surface Area

The polar surface area is another molecular property which has proved popular in the derivation of ADMET models, especially for the prediction of oral absorption and brain penetration. The polar surface area is defined as the amount of molecular surface arising from polar atoms (nitrogen and oxygen atoms together with their attached hydrogens; some definitions also include sulphur atoms and their attached hydrogens). One of the earliest uses of polar surface area in ADMET modelling employed a rather complex and time-consuming procedure for its calculation involving molecular dynamics simulations to explore the conformational space of the molecule, conformations being extracted at predefined intervals. The polar surface area is calculated for each of these structures, the value for the whole molecule being equal to the average for each of the snapshots, weighted by their energies [Palm et al. 1996]. It was subsequently demonstrated that acceptable models could be obtained using the polar surface area of a single conformation produced by a structure-generation program such as CONCORD or CORINA [Clark 1999], a much more rapid procedure. Further speed enhancements were obtained with the development of a fragment-based model wherein the polar surface area of the molecule is given by the weighted

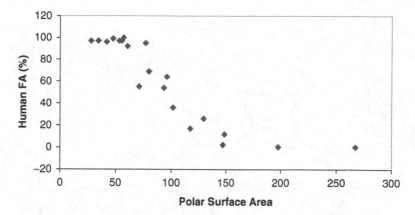

Figure 8-6. The relationship between human fractional absorption and the polar surface area [Clark 1999]. It was suggested that a cut-off around 140 Å^2 was associated with poor absorption.

sum of the contributions from each of the fragments present [Ertl et al. 2000]. This latter value can be calculated from the molecular graph and so is extremely fast, yet shows an excellent correlation with the more time-consuming methods based on the 3D structure.

One example of the use of polar surface area in deriving models is the following equation to describe Caco-2 permeability ($\log P_{app}$) [van de Waterbeemd et al. 1996]:

$$\log P_{app} = 0.008 \times MW - 0.043 \times PSA - 5.165 \tag{8.3}$$

When human absorption data were examined it was observed that a sigmoidal relationship appeared to exist between a molecule's polar surface area and its fractional absorption, suggesting that a polar surface area greater than 140 Å^2 would be associated with poor absorption [Clark 1999] (see Figure 8-6). Such a simple rule is clearly well suited to large-scale virtual screening.

4.3 Descriptors Based on 3D Fields

One family of descriptors that appears to have particular utility for the prediction of pharmacokinetic properties is the Volsurf descriptors developed by Cruciani et al. [2000]. These descriptors are based on 3D molecular fields similar to those used for the calculation of molecular similarity (as discussed in Chapter 5). They are also a crucial component of the CoMFA method for the construction of 3D-QSAR models, considered in Chapter 4. The first part of the Volsurf procedure

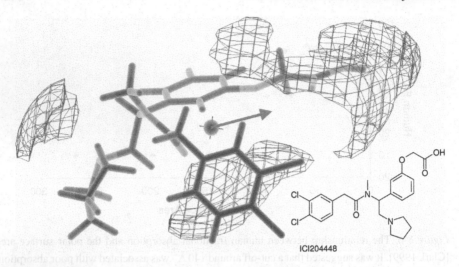

ICI204448

Figure 8-7. Schematic illustration of the calculation of an integy moment, defined as the vector from the centre of mass to the centre of the hydrophilic or hydrophobic region contoured at a particular energy level. In this case the surface shown is the contour produced using the water probe at −4.0 kcal/mol for the molecule shown.

is to compute the fields. This is done by placing each molecule into a rectangular 3D grid. A probe group is placed at each grid vertex and the interaction energy between the probe and the molecule is calculated. Various methods can be used to calculate these interaction energies; Volsurf uses the GRID program [Goodford 1985], which has been very widely used for the analysis of protein binding sites in structure-based design. Three probes are typically used: the water probe, the hydrophobic probe (known as the DRY probe within the GRID program) and the oxygen carbonyl probe. These molecular fields are then transformed into a set of molecular descriptors which quantify the molecule's overall size and shape and the balance between hydrophilicity, hydrophobicity and hydrogen bonding. They include, for example, the volumes of the 3D surfaces obtained by contouring the interaction maps with the water and hydrophobic probes at a number of different energy levels. *Integy moments* are also computed; these are similar to dipole moments and reflect the degree of imbalance between the molecule's centre of mass and the location of its hydrophilic and hydrophobic regions (Figure 8-7). The amphiphilic moment is a vector from the centre of the hydrophobic domain to the centre of the hydrophilic domain; this is believed to be related to the ability of the molecule to permeate a membrane.

Models are then built using methods such as PCA and PLS. An example is the model for predicting blood-brain barrier permeation [Crivori et al. 2000]. Training and test sets containing 110 and 120 structures, respectively, were identified from the literature. The GRID and Volsurf calculations resulted in 72 descriptors for each molecule. One of the resulting models was based on the three statistically significant principal components from the PCA. Indeed, it was found that the first principal component alone was able to provide a good discrimination between the brain-penetrating and non-penetrating compounds in the training set. This model was able to correctly classify 90% of the penetrating compounds in the test set and 65% of the non-penetrating compounds. A second model based on the two significant latent variables from the PLS proved even better, with more than 90% prediction accuracy. Two features were of particular interest from this analysis. First, examination of some of the compounds that were incorrectly predicted by the models revealed structures known to penetrate the blood-brain barrier by mechanisms other than passive permeation (e.g. via transporters) and highlighted some cases of possible mis-classification in the literature. Second, it was possible to determine the contribution of each of the descriptors to the models. This revealed that related descriptors of polarity (e.g. hydrophilic regions and hydrogen bonding) were inversely correlated with permeability whilst descriptors of hydrophobic interactions were directly correlated, though to a lesser extent.

The problem of predicting potential P450 liabilities has been tackled using a variety of approaches. These include the development of "traditional" QSAR models together with 3D pharmacophore models [de Groot and Ekins 2002] but more recent methods have taken advantage of the growing body of knowledge concerning the 3D structures of these important enzymes. A number of groups have now published x-ray structures of several P450 isoforms, in some cases with ligands bound. This information can be used for docking studies as well as more complex calculations. The structures have also been used to develop predictive models of a molecule's site(s) of metabolism by a variety of P450 enzymes. The metasite program [Cruciani et al. 2005] calculates the probability that an atom in a molecule is a potential site of metabolism as the product of two factors. The first of these is an accessibility term that represents the ability of the ligand to be positioned within the enzyme active site so that the atom is exposed towards the catalytic heme group. This accessibility term is computed using GRID molecular fields of both protein and ligand. The second contribution is a reactivity term that is computed using molecular orbital calculations and depends on the ligand 3D structure and on the specific mechanism of reaction. The result is a ranking of the probabilities, wherein the user can identify which are the most likely regions in the molecule to undergo metabolism by each of the different P450 enzymes. In the published studies the method predicted the correct site of metabolism within the top two suggestions in about 85% of the cases and in 80% of the cases the correct isoform involved was correctly predicted. However, as the authors pointed out, the methodology is not able to predict the relative rate of metabolism of a compound across the different P450 isoforms.

4.4   Toxicity Prediction

The prediction of toxic liabilities is recognised as a particularly challenging problem due in part to the multitude of possible toxicity mechanisms. It may be more feasible to focus on a single toxicological phenomenon and/or to examine just a single class of compounds (a classic example being studies of the carcinogenicity of polycyclic aromatic hydrocarbons). Nevertheless, several attempts have been made to try and predict more general toxicity. One of the most widely used methods is the DEREK system (DEREK is an acronym for Deductive Estimation of Risk from Existing Knowledge). DEREK is an expert- or rule-based system [Sanderson and Earnshaw 1991]. Central to DEREK is a set of rules that are derived from experts in the field and from the literature. These rules are based on a list of chemical substructures that have been related to specific toxic effects. The rules are combined with an "inference engine" that determines which, if any, of the rules applies to the molecule in question. The output displayed to the user indicates which of the rules apply to the molecule together with background information such as relevant literature references. If desired, new rules can be added by the user to supplement those supplied with the system.

The use within DEREK of rules derived from human experts contrasts nicely with an alternative approach that uses the CASE and MULTICASE methods. As described in Chapter 7, these automatically generate substructure-based rules from training sets of active and inactive compounds and they have been applied to a wide variety of toxicity-related measurements [Rosenkranz et al. 1999]. An additional difference between these two approaches is that whereas DEREK only identifies "activating" fragments (i.e. substructures that give rise to a toxic effect), the CASE/MULTICASE approach is also in principle able to identify both activating and deactivating fragments. A third widely used system, TOPKAT, is based on a more traditional QSAR approach using electrotopological descriptors [Gombar and Enslein 1995]. This system can also distinguish toxic and non-toxic compounds and provides tools to indicate the level of confidence in the prediction, based on the similarity to molecules in the training set and its coverage. These and other programs have been subjected to a number of evaluations to compare their performance [Durham and Pearl 2001; Greene 2002]. Most of the studies performed to date suggest that care is still required when using these systems and that the involvement and intervention of human experts is desirable. Some improvements in performance can sometimes be achieved by combining the results from more than one program in a manner reminiscent of the consensus scoring described above for docking.

5.   SUMMARY

Virtual screening methods are central to many chemoinformatics problems. Most design, selection and analysis procedures use at least one, and often several, virtual screening steps. The numbers of molecules that can be evaluated using virtual screening continues to increase with the advent of new algorithms and improved computer hardware. Nevertheless, in some areas there remain key issues, such as the reliability and accuracy of the methods for scoring the results of docking and the models for predicting ADMET properties. The lack of sufficient quantities of reliable and consistent experimental data is one of the main hindrances to the development of better models. One illuminating paper that highlights a number of such issues was the analysis by Stouch et al. [2003] of the apparent "failure" of several *in silico* ADMET models. Four models had been developed by an external software vendor. However, when tested on internal data each model was found to be deficient, for a variety of reasons. In one case, the data used to construct the model (of hERG inhibition) was taken from the literature and as a consequence derived from experiments in many different laboratories using different protocols and experimental conditions. In the second evaluation the model was built on a series of 800 marketed drugs but it was found that these occupied a different region of chemical space to that of the molecules used in the evaluation. The third model was designed to give predictions of P450 activity but not in regions of activity relevant to practical drug discovery. The fourth test was of a solubility model where there were some potential differences between the experimental protocols used in the generation of the data upon which the model was built and the evaluation data set. More importantly, only a small range of solubilities was used to build the model with a lower limit of 1 μM, which as the authors state "is, sadly, a solubility value to which only a few welcome Discovery compounds can aspire, much less exceed".

Chapter 9

COMBINATORIAL CHEMISTRY AND LIBRARY DESIGN

1. INTRODUCTION

The invention of combinatorial chemistry in the late 1980s and early 1990s led to the development of a wide variety of automated methods for chemical synthesis [Terrett 1998; Lebl 1999]. These methods range from complex robots designed for the synthesis of large combinatorial libraries to "low-tech" equipment that enables basic functions such as heating or separation to be applied to a small number of samples. The common feature of these techniques is that they enable tasks previously applied on a molecule-by-molecule basis to be applied to many molecules simultaneously, greatly increasing the rate at which new chemical entities can be made. The early work in combinatorial chemistry was driven by the needs of the pharmaceutical industry, but there is now much interest in automated synthesis techniques in other areas of chemistry such as materials science [Xiang et al. 1995; Danielson et al. 1997].

The origins of combinatorial chemistry lie in the use of solid supports for peptide synthesis. By coupling the growing peptide to a solid support such as a polystyrene bead it is possible to use excess reagents and so ensure that the reaction proceeds to completion. Any excess reagent is simply washed away. In the original applications of solid-phase chemistry to peptide synthesis the goal was generally the synthesis of a single molecular target. A key breakthrough was the recognition that this methodology could be used to generate large numbers of molecules using a scheme known as *split–mix*. This is illustrated in Figure 9-1; it starts with a set of reagents (which we may also refer to as "monomers"), each of which is coupled to the solid support. These are then all mixed together and divided into equal-sized aliquots for reaction with the second reagent. The products from this reaction are then mixed together and divided for reaction with the third reagent and so on. If the numbers of reagents at each step are n_1, n_2, n_3, etc. then the total number of molecules produced is the product $n_1 n_2 n_3$. The size of the library thus increases exponentially with the number of reagents – hence the use of the term "combinatorial".

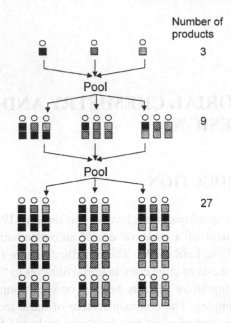

Figure 9-1. Illustration of the split–mix approach to combinatorial synthesis, using three sets each containing three monomers. (Adapted in part from Leach 2001.)

The original split–mix method is capable of producing extremely large libraries, but it does suffer from some drawbacks. A particular limitation is that due to the various mixing stages the identity of the product on each bead is unknown (except for the final reagent). It is important to note, however, that each bead contains just one discrete chemical entity. Much progress has subsequently been made in the technology for automated synthesis and purification since the first reports were published. These developments have enabled many of the limitations of the early combinatorial techniques to be overcome, making automated synthesis methods commonplace in both industrial and academic laboratories

In this chapter we will consider both theoretical and practical facets of library design. Some of these are unique to libraries, such as enumeration and the combinatorial constraints imposed on monomer selection methods. Other aspects, such as virtual screening, have been considered in detail in earlier chapters. A key area that continues to be debated concerns the balance between diversity and focus when designing libraries. We conclude with some examples taken from the literature to indicate the range of applications of library design.

2. DIVERSE AND FOCUSSED LIBRARIES

The initial emphasis in combinatorial chemistry was on the synthesis of as many "diverse" compounds as possible, in the expectation that a larger number of molecules would inevitably lead to more hits in the biological assays. This was not observed in practice; "historical" compounds built up over many years generally performed much better than the new libraries. Moreover, the hits produced from combinatorial chemistry often had properties that made them unsuitable as lead compounds (e.g. they were too large, too insoluble, or contained inappropriate functional groups). This is illustrated in Figure 9-2 which shows the molecular weight profile of an aminothiazole library designed to be "diverse" superimposed on the corresponding profile of compounds in the WDI. It can be seen that the molecules in the library tend to have higher molecular weights than is seen in known drug molecules.

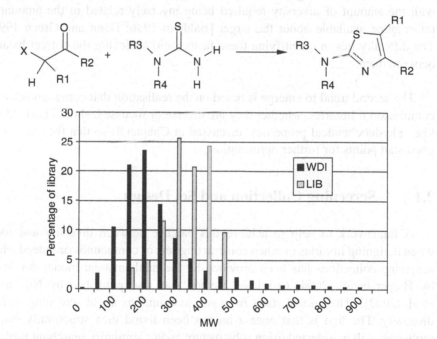

Figure 9-2. Comparison of the molecular weight distributions of an aminothiazole library designed to be diverse with that for the WDI. The library of 450 products was constructed from 15 α-bromoketones and 30 thioureas selected from a virtual library of 12,850 potential products (74 α-bromoketones × 174 thioureas). Diversity was assessed using the MaxSum score based on Daylight fingerprints and the cosine coefficient (see Chapter 5 for more details of this diversity measure). The distribution for the library is shifted significantly towards the inclusion of higher molecular weight species.

Since the early days of combinatorial chemistry, two trends have emerged. The first is towards the design of smaller, more focussed libraries that incorporate as much information about the therapeutic target as possible. Of course, the amount of information that is available will vary from one drug discovery project to another. At one extreme, when the 3D structure of the target protein is available, combinatorial libraries (and screening libraries in general) may be designed to fit the active site using docking methods. If some active compounds are known then libraries may be designed using pharmacophoric approaches, 2D similarity or physicochemical properties. A recent trend is to focus libraries on a family of targets, such as proteases or kinases, in which case the design should incorporate properties that are known to be important for all targets in the family. When little is known about the target then more diverse libraries may be relevant. Such libraries are sometimes referred to as primary screening libraries and are designed to give a broad coverage of chemistry space so that they can be screened against a range of structural targets. In general, a balance between diversity and focus is needed, with the amount of diversity required being inversely related to the amount of information available about the target [Baldwin 1996; Hann and Green 1999]. The difficulty lies in quantifying these factors and achieving the correct balance between them.

The second trend to emerge is based on the realisation that compounds within combinatorial libraries, whether they are diverse or focussed, should have "drug-like" physicochemical properties (discussed in Chapter 8) so that they constitute good start points for further optimisation.

2.1   Screening Collection and Set Design

A framework to help to achieve this balance between diversity and focus when designing libraries or when constructing sets of compounds or indeed whole screening collections has been provided by the mathematical model developed by Harper et al. [2004] (a related method was also published by Nilakantan et al. [2002]). This model starts from two assumptions about screening in drug discovery. The first is that once a hit has been found then structurally similar molecules will be selected using substructure and/or similarity searching methods for further testing in order to provide evidence of a structure–activity relationship. The second assumption is that the overall goal is to find multiple lead series from the screen.

The chemical space is assumed to be clustered or partitioned in some way into structural clusters. Two parameters are introduced into the model. These are the probability that a randomly selected molecule from some cluster i will test active

(α_i) and the probability that the cluster will contain a lead molecule (π_i). A lead molecule is more than just a "hit"; it is a compound with the correct combination of activity, physicochemical, ADMET properties and novelty to provide a good chance of becoming a marketed drug. Based on these two parameters it is possible to derive a number of useful quantities, including the expected number of lead-containing clusters in which leads are found, E:

$$E = \sum_{i=1}^{p} \pi_i \left[1 - (1 - \alpha_i)^{N_i} \right]$$ (9.1)

where N_i compounds are screened from p clusters. In addition the probability of finding no leads at all (a quantity that should be as close to zero as possible) is:

$$P_{\text{noleads}} = \prod_{i=1}^{p} \left[1 - \pi_i \left(1 - (1 - \alpha_i)^{N_i} \right) \right]$$ (9.2)

The consequences of different ways to construct a compound collection can then be derived. For example, if one uses values for π_i of 0.00001 (a value typical for a set of "diverse" compounds screened against tractable drug targets) and α_i of 0.3 [Martin et al. 2002] then for a set of 1 million compounds there is a probability of finding no leads greater than 0.9 if the set comprises clusters of size 100. However, if the collection has two compounds per cluster then the probability of finding no leads drops to less than 0.08. The model enables the optimal design of various types of compound sets to be determined, be they screening collections, focussed sets or combinatorial libraries. Moreover, the model enables a diversity score to be derived that conforms to all but one of the criteria proposed by Waldman et al. [2000].

3.   LIBRARY ENUMERATION

The analysis of a combinatorial library obviously requires computer representations of the molecules contained within it. Given the large numbers of compounds that are usually involved, it is not practical to input each product molecule individually. *Enumeration* refers to the process by which the molecular graphs (e.g. connection tables or SMILES strings) of the product molecules are generated automatically from lists of reagents. Enumeration may be required for both virtual libraries and for libraries that have actually been synthesised.

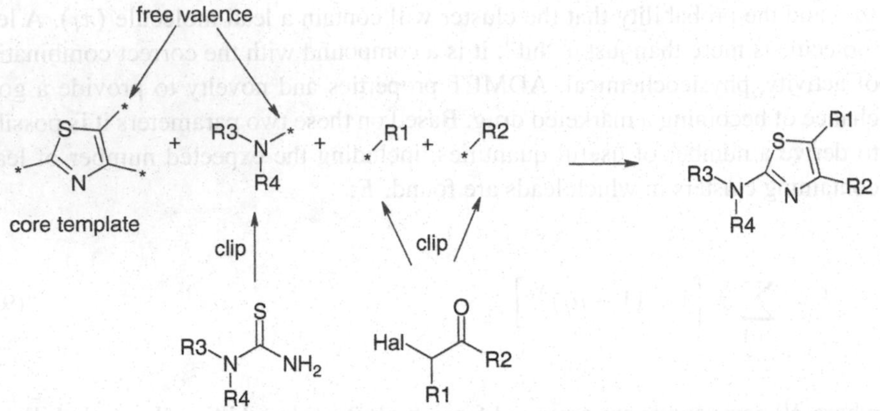

Figure 9-3. Illustration of the fragment marking approach to library enumeration.

There are two general approaches to enumeration. The first of these is called *fragment marking* and considers the library to be composed of a central core template with one or more *R groups*. The core template must be common to all members of the library; different product structures correspond to different combinations of R groups. In order to enumerate the library, bonds are created between the template and the appropriate R groups. The first task is to generate the R group structures from the corresponding reagents. This is most easily achieved by replacing the appropriate reactive group by a "free valence" to give a "clipped" reagent. This procedure is illustrated in Figure 9-3 for the enumeration of aminothiazoles from α-haloketones and thioureas in which a central core template corresponding to the thiazole ring is constructed with three attachment points. The three sets of clipped fragments would then be generated from the reagent structures for combination with the core template to produce complete product structures.

The alternative to fragment marking is the reaction transform approach. Central to this method is the transform itself, which is a computer-readable representation of the reaction mechanism. The enumeration program takes as input the reagent structures themselves and applies the transform to generate the products. The reaction transform approach thus more closely mimics the actual synthetic process. As can be seen from the example in Figure 9-4 it is necessary to ensure that the transform is defined so that the appropriate atoms are matched between reagents and products (i.e. according to the reaction mechanism). This may be achieved using *atom mapping*, where pairs of corresponding atoms in the reagents and products are labelled, as shown (see also the discussion on reaction databases in Chapter 1).

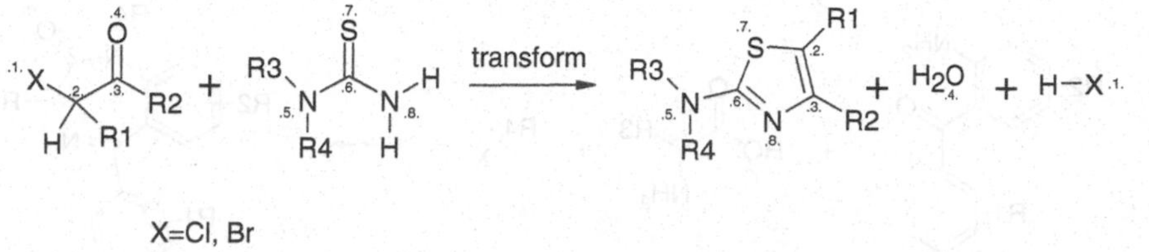

X=Cl, Br

Figure 9-4. Illustration of library enumeration using the reaction transform approach. The integers give the mapping between atoms in the reactants and the products.

Both the fragment marking and the reaction transform approaches have advantages and disadvantages. Fragment marking generally enables the enumeration to be performed very quickly once the core template and the R group fragments have been defined. It is particularly useful when the library can be naturally defined as a "core plus R group" and when the reagents can be easily clipped using some form of automated procedure. In other cases, however, it can be difficult to define a common core and it may prove difficult to generate the fragments automatically. Certain types of reactions (e.g. Diels–Alder reactions) may not be handled properly by a simple fragment marker, giving rise to extraneous and incorrect structures. The reaction transform method by contrast more closely corresponds to the actual chemical synthesis and so it may prove more intuitive to bench scientists, especially if delivered via an easy-to-use interface such as the web [Leach et al. 1999]. Some expertise is required to compose efficient transforms, but once done they can be reused many times without further alteration.

A more recent development is the use of Markush-based approaches to enumeration, building upon research into the representation and search of the Markush structures found in patents (see Chapter 1). This approach is ideally suited to combinatorial libraries, particularly where a common core can be identified. The Markush approach [Downs and Barnard 1997] recognises that certain subsets of the product structures in the library may have features in common (e.g. a subset of the R groups at a particular position may contain a benzoic acid). The method constructs a compact computational representation of the library from which the structures of the enumerated library can be generated very rapidly.

4. COMBINATORIAL LIBRARY DESIGN STRATEGIES

In most combinatorial reaction schemes there are many more monomers available than can be handled in practice. For example, consider a benzodiazepine

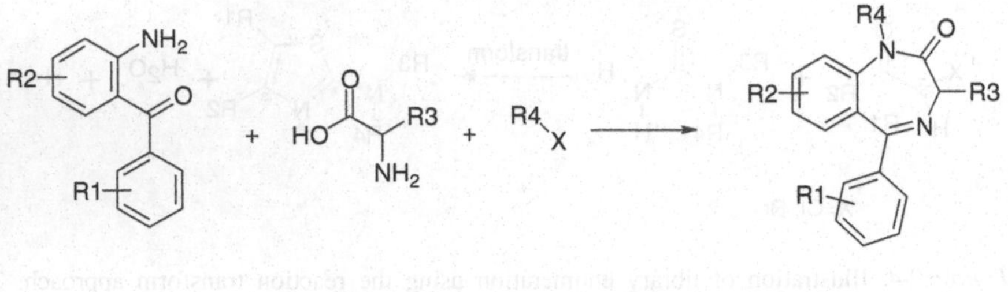

Figure 9-5. Synthesis of benzodiazepines.

library constructed from three monomer pools, as shown in Figure 9-5. It is easy to find many monomers for each position of variability by searching a database of available reagents such as the ACD [ACD]. The virtual libraries in such cases can easily contain millions if not billions of products; much larger than can actually be synthesised in practice. Thus, a key issue in combinatorial library design is the selection of monomers such that the resultant library is of a manageable size and has the desired properties, whether it is to be diverse or focussed or should exhibit a balance between the two.

The two main strategies for combinatorial library design are known as *monomer-based selection* and *product-based selection*.

4.1   Monomer-Based Selection

In monomer-based selection optimised subsets of monomers are selected without consideration of the products that will result. Consider a hypothetical three-component library with 100 monomers available at each position of variability, where the aim is to synthesise a diverse $10 \times 10 \times 10$ combinatorial library. In monomer-based selection this would involve selecting the 10 most diverse monomers from each set of monomers. In general, there are

$$\frac{N!}{n!(N-n)!} \tag{9.3}$$

subsets of size n contained within a larger set of N compounds. For example, there are more than 10^{13} different subsets of size 10 from a pool of 100 monomers. It is not possible to examine all of these. We discussed the subset selection problem in Chapter 6 in the context of selecting compounds for screening where the techniques of dissimilarity-based compound selection, clustering and partitioning were introduced, together with related optimisation methods. Any of these techniques can also be used in the monomer-based selection approach.

An early example of monomer-based design is described by Martin and colleagues and used experimental design. Diverse subsets of monomers were selected for the synthesis of peptoid libraries (peptoids are synthetic oligomers with a peptide backbone but with the side chain attached to the nitrogen atom rather than the alpha-carbon atom) [Martin et al. 1995]. The diversity measure included features such as lipophilicity, shape and chemical functionality and uses a variety of descriptors including ClogP, topological indices, Daylight fingerprints and atom layer properties based on receptor recognition descriptors. Diverse subsets of monomers were selected using D-optimal design.

4.2   Product-Based Selection

In product-based selection, the properties of the resulting product molecules are taken into account when selecting the monomers. Having enumerated the virtual library any of the subset selection methods introduced in Chapter 6 could then in principle be applied. This process is generally referred to as *cherry-picking* but it is synthetically inefficient insofar as combinatorial synthesis is concerned. Synthetic efficiency is maximised by taking the *combinatorial constraint* into account and selecting a combinatorial subset such that every reagent selected at each point of variation reacts with every other reagent selected at the other positions.

Product-based selection is much more computationally demanding than monomer-based selection. The number of combinatorial subsets in this case is given by the following equation:

$$\prod_{i=1}^{R} \frac{N_i!}{n_i!(N_i - n_i)!} \tag{9.4}$$

where R is the number of positions of variability and there are n_i monomers to be selected from a possible N_i at each substitution position. Thus, there are almost 10^{40} different $10 \times 10 \times 10$ libraries that could be synthesised from a $100 \times 100 \times 100$ virtual library. The selection of combinatorial subsets has been tackled using optimisation techniques such as simulated annealing and genetic algorithms as will be discussed in detail below.

Despite the greater computational complexity of performing product-based selection compared to monomer-based selection it can be a more effective method when the aim is to optimise the properties of a library as a whole, such as diversity or the distribution of physicochemical properties. For example, it has been shown that more diverse libraries result if selection is performed in product-space rather than in monomer-space [Gillet et al. 1997; Jamois et al. 2000]. In addition, product-based selection is usually more appropriate for focussed libraries which require consideration of the properties of the product structures.

5. APPROACHES TO PRODUCT-BASED LIBRARY DESIGN

A general strategy for product-based library design involves the following three steps. First, lists of potential reagents are identified (e.g. by searching relevant databases), filtered as appropriate, and the virtual library is enumerated. In the second step the virtual library is subjected to virtual screening to evaluate and score each of the structures. In the third stage the reagents to be used in the actual library for synthesis are selected using the results from the virtual screening together with any additional criteria such as the degree of structural diversity required, or the degree of similarity/dissimilarity needed to existing collections. This three-stage process is illustrated in Figure 9-6.

The first two stages of monomer identification and enumeration, and virtual screening, have been described above and in Chapter 8. It is important to note that it may be possible to reduce significantly the size of the virtual library by

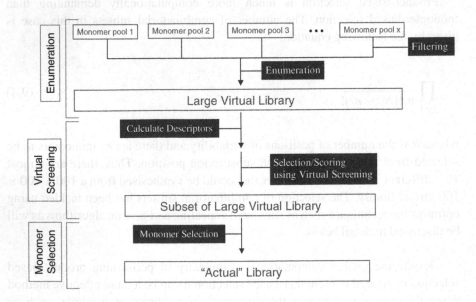

Figure 9-6. Three-stage procedure for library design.

eliminating from consideration monomers that can be unambiguously identified as being inappropriate. For example, monomers that contain functionality known to be incompatible with the proposed reaction scheme can be eliminated, as can monomers that are unavailable or cannot be purchased in time. This is the initial "filtering" step in Figure 9-6.

The final, monomer selection stage is typically implemented using optimisation techniques such as GAs or simulated annealing [Agrafiotis 1997; Brown et al. 2000; Zheng et al. 1998, 2000]. By way of example, the SELECT program [Gillet et al. 1999] is based on a GA in which each chromosome encodes one possible combinatorial subset. Assume a two component combinatorial synthesis in which n_A of a possible N_A first monomers are to be reacted with n_B of a possible N_B second monomers. The chromosome of the GA thus contains $n_A + n_B$ elements, each position specifying one possible monomer. The fitness function quantifies the "goodness" of the combinatorial subset encoded in the chromosome and the GA evolves new potential subsets in an attempt to maximise this quantity.

In some cases the virtual library is too large to allow full enumeration and descriptor calculation, making product-based combinatorial subset selection unfeasible. A number of methods have been proposed to try to overcome this problem. For example, the Markush-based enumeration strategy described above enables certain types of product descriptors to be calculated very rapidly without actually constructing the products themselves [Barnard et al. 2000]. In the first phase, descriptors are calculated for each of the monomers in the library. In the second phase the monomer descriptors are combined together to generate the descriptor for the product. Such an approach is applicable to any properties that are additive, such as molecular weight, counts of features such as donors and acceptors, calculated $\log P$ and even 2D fingerprints. An alternative approach is to use random sampling techniques to derive a statistical model of the property under consideration; in addition to the estimation of molecular properties such an approach may be used for monomer selection and to prioritise different library scaffolds [Beroza et al. 2000].

Alternative approaches to product-based library design have been developed that do not require enumeration of the entire virtual library. These methods have been termed *molecule-based methods* to distinguish them from library-based methods [Sheridan et al. 2000] and they are appropriate for the design of targeted or focussed libraries. For example, the method developed by Sheridan and Kearsley [1995] uses a GA in which each chromosome encodes a product molecule (rather than a combinatorial library). The fitness function measures the similarity of the molecule encoded by a chromosome to a target molecule.

When the GA has terminated the entire population of chromosomes is analysed to identify monomers that occur frequently across all the molecules in the population. The frequently occurring monomers could then be used to construct a combinatorial library. The method was tested on a tripeptoid library with three positions of variability; 2,507 amines were available for two of the substitution positions and 3,312 for the third position. This gives a virtual library of approximately 20 billion potential tripeptoids. The method was able to find molecules that were very similar to given target molecules after exploring only a very small fraction of the total search space.

The molecule-based method is a relatively fast procedure, especially when optimisation is based on 2D properties, since the fitness function involves a pairwise molecular comparison rather than the analysis of an entire library, as is the case in library-based methods. In these approaches, however, there is no guarantee that building libraries from frequently occurring monomers will result in optimised libraries, nor is it possible to optimise properties of the library as a whole [Bravi et al. 2000].

6.   MULTIOBJECTIVE LIBRARY DESIGN

The approaches to library design methods described so far have generally been concerned with the optimisation of a single objective, such as diversity or similarity to a target. However, it is usually desirable to optimise multiple properties simultaneously. For example, in addition to achieving a balance between diversity and focus, it may be desired to design libraries with "drug-like" physicochemical properties from readily available, inexpensive reagents and so on. Such a situation is referred to as *multiobjective library design*.

Most approaches to multiobjective library design combine the different properties via a weighted-sum fitness function. For example, in the SELECT program different objectives can be specified by the user who also assigns their relative weights. To optimise the diversity whilst achieving a particular molecular weight profile one may use a fitness function of the following form:

$$f = w_1(1 - d) + w_2 \Delta MW \tag{9.5}$$

where d is diversity (i.e. minimising $(1 - d)$ is equivalent to maximising diversity) and ΔMW is the difference between the molecular weight profile of the library and that which is desired (e.g. a profile derived from a set of known active compounds). The aim is therefore to minimise both terms. When this function is used to design an aminothiazole library, the molecular weight profile of the resulting library is

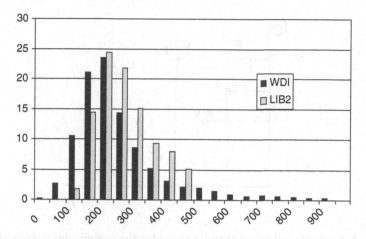

Figure 9-7. Molecular weight distribution for a 15 × 30 aminothiazole library designed to be both diverse and to achieve a molecular weight distribution that matches that of the WDI.

shown in Figure 9-7. This should be compared with the profile in Figure 9-2 which is for an equivalent library designed using the diversity function alone. It is clear that the library designed with the multiobjective function has a much more "drug-like" profile than is the case for the library optimised on diversity alone. The improved profile is achieved at the expense of some diversity but this may be deemed an acceptable compromise, especially as it corresponds to the removal of "non-drug-like" compounds.

The weighted-sum approach is easily extended to include additional physicochemical properties such as the distribution of ClogP, and any other calculable properties such as cost. A similar approach to multiobjective library design has been implemented in several other library design programs [Good and Lewis 1997; Brown et al. 2000; Agrafiotis 2002]. Tailoring combinatorial libraries in this way should increase the chances of finding useful lead compounds.

6.1 Multiobjective Library Design Using a MOGA

Despite the obvious benefits of optimising libraries over several properties simultaneously, in practice it can be difficult to find an appropriate compromise via a weighted-sum fitness function. Often, several trial-and-error runs are required. A further disadvantage is that each run results in a single solution, when in fact an entire family of solutions exists.

A way to overcome these limitations is to use a *Multiobjective Genetic Algorithm* (MOGA) [Fonseca and Fleming 1993; Coello et al. 2002]. In a MOGA,

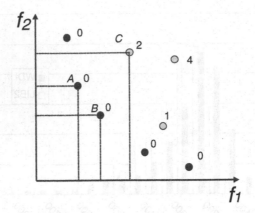

Figure 9-8. The Pareto-optimal solutions for a two-objective minimisation problem are shown in black. Solutions *A* and *B* are both non-dominated and so they are given a rank of zero. Solution *C* is dominated by both *A* and *B* and hence it is given a rank of two.

each objective is handled independently without the need for summation and the use of relative weights. A MOGA attempts to identify a *Pareto-optimal* set of solutions where a Pareto-optimal solution, also called a *non-dominated* solution, is one for which no other solution exists in the population that is better in all objectives. The weighted-sum fitness function used in a traditional GA is replaced by fitness based on dominance; an individual is given a rank according to the number of individuals in the population by which it is dominated as shown in Figure 9-8. The fitness of an individual is then calculated such that all individuals with the same rank have the same fitness value, with the least dominated individuals being preferred. The result is a family of solutions, each representing a different compromise between the individual objectives. These solutions define the Pareto surface, an example of which is shown in Figure 9-9 for our aminothiazole library which has been optimised using a MOGA on both diversity and on the molecular weight profile. This method allows the relationships between the objectives to be explored in a single run, thereby removing the trial-and-error approach that is required with a weighted sum. The library designer can therefore make an informed choice about any compromises that may need to be made between the objectives [Gillet et al. 2002a, b].

7.   PRACTICAL EXAMPLES OF LIBRARY DESIGN

In this final section we will give two examples of library design, taken from the literature. The examples have been chosen to illustrate some of the aspects of current library design that have been discussed above.

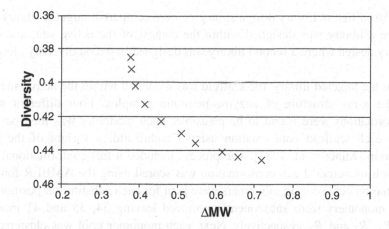

Figure 9-9. Family of solutions obtained with a MOGA algorithm designed to investigate the balance between diversity and molecular weight profile for aminothiazole libraries. Each symbol represents a different potential library.

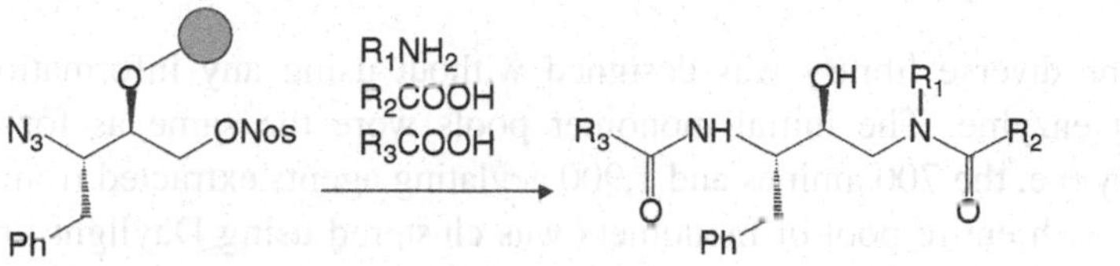

Figure 9-10. The scaffold used in the design of inhibitors of Cathepsin D.

7.1 Structure-Based Library Design

Our first example concerns a library designed to find inhibitors of Cathepsin D, an aspartyl protease that has been implicated in tumour metastasis in breast cancer, melanoma metastasis and Alzheimer's disease [Kick et al. 1997]. At the outset of the study no potent non-peptidic inhibitors were known. Libraries were designed based on a known aspartyl-protease transition-state mimetic with three positions of diversity as shown in Figure 9-10.

Monomers for each substitution position were extracted from the ACD [ACD] and were filtered so that monomers with molecular weight greater than 275 and compounds that were incompatible with the synthesis were eliminated. This resulted in 700 amines and approximately 1,900 acylating and sulphonylating agents (carboxylic acids, sulphonyl chlorides and isocyanates), representing a virtual library of more than 1 billion possible compounds.

Two different library design strategies were compared: targeted library design where a library was designed within the context of the active site; and diverse library design where a second library was designed based on diversity alone.

In the targeted library, the scaffold was modelled within the active site, based on the x-ray structure of enzyme–pepstatin complex. Four different scaffold conformations were found to be plausible. Each monomer was then positioned onto each scaffold conformation using CombiBuild, a variant of the DOCK program [Kuntz et al. 1982]. The process included a full conformational search of each monomer. Each conformation was scored using the AMBER force field and the 50 best-scoring monomers were kept for each substitution position. High cost monomers were subsequently removed leaving 34, 35 and 41 monomers for R_1, R_2 and R_3, respectively. Next, each monomer pool was clustered using Daylight fingerprints as descriptors and an hierarchical clustering method. Ten monomers were selected for each monomer pool from the resulting clusters. The resulting $10 \times 10 \times 10$ library was then synthesised and screened for activity against Cathepsin D.

The diverse library was designed without using any information about the target enzyme. The initial monomer pools were the same as for the targeted library (i.e. the 700 amines and 1,900 acylating agents extracted from ACD). This time, each entire pool of monomers was clustered using Daylight fingerprints as descriptors and Jarvis–Patrick clustering. Ten monomers were selected from each pool to maximise diversity; each monomer was taken from a unique cluster with additional factors such as size and synthetic accessibility also being taken into account. The diverse $10 \times 10 \times 10$ library was synthesized and screened for activity against Cathepsin D.

The targeted library resulted in 67 compounds at 1 μM activity, 23 compounds at 333 nM and 7 compounds at 100 nM activity. The diverse library, which contained a greater range of functionality, gave 26 compounds at 1 μM, 3 compounds at 333 nM and 1 compound at 100 nM activity. Thus, the targeted library proved to be much more effective at identifying active compounds than the diverse library. This study clearly demonstrates the power of integrating combinatorial chemistry methods with structure-based design techniques, an area of significant current interest [Kubinyi 1998b; Stahl 2000; Böhm and Stahl 2000].

7.2   Library Design in Lead Optimisation

Our second example concerns the use of solid-phase synthesis to identify potent and orally bioavailable P38 MAP kinase inhibitors [McKenna et al.

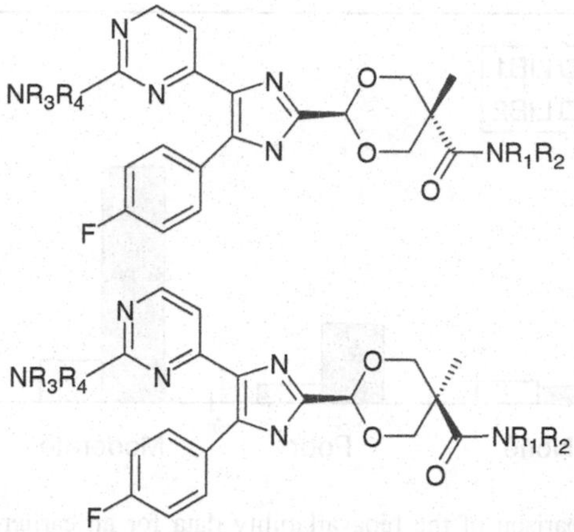

Figure 9-11. Structure of *trans* and *cis* isomers of the pyrimidinoimidazole-based library with two points of diversity.

2002]. Previous work had identified a series of imidazolyl-based cyclic acetals as inhibitors of this enzyme [McLay et al. 2001]; the new goal was the synthesis of a library with two points of diversity that would lead to potent inhibitors with greater bioavailability.

The generic structure of the library is shown in Figure 9-11; the points of variation arise from the use of two sets of amines, R_1R_2NH and R_3R_4NH, the latter forming an amide linkage to the 1,3-dioxane ring and the former undergoing nucleophilic attack at the 2-position of the pyrimidine ring. These compounds exist as *cis* or *trans* isomers according to the relative stereochemistry about the dioxane ring. Both isomers were synthesised as separate libraries, with the same set of monomers being used for each.

The library design was performed using a Monte Carlo selection algorithm [Pickett et al. 2000]. The initial virtual library contained 1,485 product structures (45 R_1R_2NH and 33 R_3R_4NH). This enumerated library was filtered to eliminate compounds predicted to have poor bioavailability properties using two virtual screens. The first filter required the polar surface area to be less than 140 $Å^2$. The second filter was a modified "rule of five" (ClogP < 6.2, molecular weight < 550, number of H-bond donors < 5, number of H-bond acceptors < 12). These modifications were derived from previous experience with this series of compounds. These two filters reduced the initial library of 1,485 compounds to 770 acceptable structures.

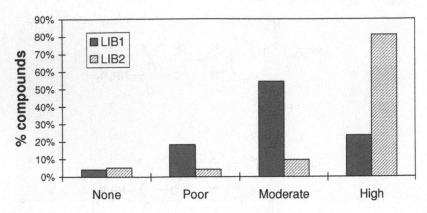

Figure 9-12. A comparison of the bioavailability data for an earlier library (LIB1) and that discussed in the text (LIB2) showing the much higher proportion of compounds having high bioavailability in the latter.

The desire was to construct a 20×20 combinatorial array. However, no such subset existed within the 770 compounds such that all 400 compounds would pass the filters. Due to the particular solid-phase synthetic route chosen, it was considered important to select as close to 20 R_1R_2NH as possible and that the number of R_3R_4NH should be equal to or greater than 20. The final library would therefore contain at least 400 compounds for each of the two isomers and be as close as possible to a combinatorial solution. It was also desired to include some of the known active monomers. These requirements were combined with appropriate weighting factors into a fitness function used by the library design algorithm which suggested 21 R_1R_2NH and 24 R_3R_4NH. Of the 504 possible compounds 449 (i.e. 89%) satisfied the two bioavailability filters. A particular aspect of the design algorithm was its ability to suggest near-combinatorial solutions such that not every combination of R_1R_2NH and R_3R_4NH was actually synthesised.

Many active compounds were identified from this library including one that was 300 times more potent in the enzyme assay than the original candidate structure. Perhaps more striking were the bioavailability results. This can be seen in Figure 9-12 which compares the bioavailability data (measured as Caco-2 permeability) for this library and one designed previously. The monomers for the earlier library had been chosen by a manual procedure with only a simple constraint on the molecular weight (MW < 550). The superior performance of the second library, designed specifically to contain molecules with improved ADME properties, is clear.

8.   SUMMARY

The rapid pace at which new experimental technologies for automated chemical synthesis have been introduced has been matched by developments in computational methods for library design. Indeed, consideration of design factors is central to the successful application of the technology. An example of the influence of such factors is the trend from large "diverse" libraries towards smaller, more focussed designs.

Many of the techniques that have been described in this book are important in library design. Thus, substructure searching of 2D databases can identify potential reagents, and reaction databases are a source of potential synthetic routes. An ever-increasing range of virtual screening approaches is used for the evaluation of virtual libraries, ranging from substructure-based filters and methods based on 2D similarity through 3D-pharmacophore approaches to high-throughput protein–ligand docking. It is increasingly desired to take ADMET properties into account. Finally, it is often necessary to achieve an appropriate balance between many different factors when making the selection of reagents for the actual synthesis, not least that between the degrees of "diversity" and "focus".

Library design has also posed new challenges. Some of these arise because the number of novel chemical structures that could be made is so enormous, leading for example to the development of new methods for library enumeration and virtual screening. The desire to take into account many different factors when designing a library has resulted in new approaches to multiobjective design problems. The development of combinatorial chemistry is one of the main reasons for the introduction of chemoinformatics as a distinct discipline; just as combinatorial and other automated chemistry techniques are now widely employed so we hope that this book will help to widen the knowledge, understanding and appreciation of chemoinformatics within the scientific community.

APPENDIX 1. MATRICES, EIGENVECTORS AND EIGENVALUES

Matrices are widely used in many areas of science and computation. Here we provide a brief summary of some of the key features relevant to the methods described in this book.

An $m \times n$ matrix is a rectangular array of quantities with m rows and n columns. Matrices can be added and subtracted if they have the same dimensions. For example:

$$\begin{pmatrix} 1 & -2 \\ 3 & 0 \end{pmatrix} + \begin{pmatrix} 4 & 1 \\ 2 & -3 \end{pmatrix} = \begin{pmatrix} 5 & -1 \\ 5 & -3 \end{pmatrix} \tag{A1.1}$$

$$\begin{pmatrix} 1 & 2 & 3 \\ 4 & 0 & 1 \end{pmatrix} - \begin{pmatrix} 1 & 4 & 0 \\ 2 & -3 & 2 \end{pmatrix} = \begin{pmatrix} 0 & -2 & 3 \\ 2 & 3 & -1 \end{pmatrix} \tag{A1.2}$$

To multiply two matrices $(\mathbf{AB})$ the number of columns in $\mathbf{A}$ must be equal to the number of rows in $\mathbf{B}$. If $\mathbf{A}$ is an $m \times n$ matrix and $\mathbf{B}$ is an $n \times o$ matrix the product $\mathbf{AB}$ is an $m \times o$ matrix. Each element (i, j) in the product matrix is obtained by multiplying each of the n values in the ith row of $\mathbf{A}$ by each of the n values in the jth column of $\mathbf{B}$. For example:

$$\begin{pmatrix} 1 & -2 \\ 3 & 0 \end{pmatrix} \begin{pmatrix} 4 & 1 \\ 2 & -3 \end{pmatrix} = \begin{pmatrix} (1 \times 4) + (-2 \times 2) & (1 \times 1) + (-2 \times -3) \\ (3 \times 4) + (0 \times 2) & (3 \times 1) + (0 \times -3) \end{pmatrix} \tag{A1.3}$$

$$= \begin{pmatrix} 0 & 7 \\ 12 & 3 \end{pmatrix} \tag{A1.4}$$

A *square matrix* has the same number of rows and columns. A *diagonal matrix* is a special type of square matrix in which every element is zero except for those on the diagonal. The *unit* or *identity matrix* $\mathbf{I}$ is a diagonal matrix in which all of the diagonal elements are equal to one. If the elements of a square matrix are mirror images above and below the diagonal then the matrix is described as *symmetric*.

The *transpose* of a matrix, $\mathbf{A}^{\mathrm{T}}$, is the matrix obtained by exchanging its rows and columns. Thus the transpose of an $m \times n$ matrix is an $n \times m$ matrix:

$$\begin{pmatrix} 1 & 2 & 3 \\ 4 & 0 & 1 \end{pmatrix}^{T} = \begin{pmatrix} 1 & 4 \\ 2 & 0 \\ 3 & 1 \end{pmatrix} \tag{A1.5}$$

The *determinants* of 2×2 and 3×3 matrices are calculated as follows:

$$\begin{vmatrix} a & b \\ c & d \end{vmatrix} = (a \times d) - (b \times c) \tag{A1.6}$$

$$\begin{vmatrix} a & b & c \\ d & e & f \\ g & h & i \end{vmatrix} = a \begin{vmatrix} e & f \\ h & i \end{vmatrix} - b \begin{vmatrix} d & f \\ g & i \end{vmatrix} + c \begin{vmatrix} d & e \\ g & h \end{vmatrix} \tag{A1.7}$$

$$= a\,(ei - hf) - b\,(di - fg) + c\,(dh - eg)$$

For example:

$$\begin{vmatrix} 1 & -2 \\ 3 & 0 \end{vmatrix} = 6; \quad \begin{vmatrix} 4 & 2 & -2 \\ 2 & 5 & 0 \\ -2 & 0 & 3 \end{vmatrix} = 28 \tag{A1.8}$$

The 3×3 determinant is thus formulated as a combination of determinants of a series of three 2×2 matrices, obtained by deleting the relevant row (i) and column (j). The determinant of each of these 2×2 matrices is then multiplied by $(-1)^{i+j}$. The determinants of higher-order matrices ($4 \times 4, 5 \times 5$, etc.) are obtained by extensions of this scheme; thus the determinant of a 4×4 matrix is initially written in terms of a series of 3×3 matrices which in turn are decomposed to a set of 2×2 matrices.

The *eigenvalues* and *eigenvectors* of a matrix are frequently required; an eigenvector is one-column matrix $\mathbf{x}$ such that:

$$\mathbf{Ax} = \lambda\mathbf{x} \tag{A1.9}$$

where λ is the associated eigenvalue. This equation can be reformulated as follows:

$$\mathbf{Ax} = \lambda\mathbf{x} \Rightarrow \mathbf{Ax} = \lambda\mathbf{xI} \Rightarrow (\mathbf{A} - \lambda\mathbf{I})\,\mathbf{x} = 0 \tag{A1.10}$$

For there to be a non-trivial solution (i.e. $\mathbf{x} \neq 0$) the determinant $|\mathbf{A}-\lambda\mathbf{I}|$ needs to equal zero. For small matrices the determinant can be expanded to give a polynomial equation in λ which can then be solved. For example, to find the eigenvalues and eigenvectors of the 3×3 matrix above:

$$\begin{vmatrix} 4-\lambda & 2 & -2 \\ 2 & 5-\lambda & 0 \\ -2 & 0 & 3-\lambda \end{vmatrix} = 0 \Rightarrow -\lambda^3 + 12\lambda^2 - 39\lambda + 28 = 0 \qquad (\text{A1.11})$$

This equation has solutions $\lambda = 1, \lambda = 4$ and $\lambda = 7$ with the corresponding eigenvectors being:

$$\lambda = 1 : \mathbf{x} = \begin{pmatrix} 2/3 \\ -1/3 \\ 2/3 \end{pmatrix} ; \lambda = 4 : \mathbf{x} = \begin{pmatrix} -1/3 \\ 2/3 \\ 2/3 \end{pmatrix} ; \lambda = 4 : \mathbf{x} = \begin{pmatrix} 2/3 \\ 2/3 \\ -1/3 \end{pmatrix} (\text{A1.12})$$

Here we have written the eigenvectors as vectors of unit length although any multiple of these would also be a solution.

For the larger matrices more typically encountered in "real" problems more efficient computational methods are used to solve the eigenvalue problem [Press et al. 1993].

APPENDIX 2. CONFORMATION, ENERGY CALCULATIONS AND ENERGY SURFACES

The conformations of a molecule are the 3D structures that it can adopt. Conformational analysis is the study of the conformations of a molecule and their influence on its properties. Different conformations of a molecule arise from changes in bond lengths, bond angles and *torsion angles* (defined in Figure A2-1). Rotations about single bonds usually give rise to the most significant changes in conformation and also require the least energy. For this reason it is common when performing a conformational analysis to keep the bond lengths and angles fixed and to vary just the rotatable bonds.

Different conformations have different energies. Such energies can be calculated using a variety of computational chemistry techniques, of which the two most common methods are *quantum mechanics* and *molecular mechanics*. In quantum mechanics the Schrödinger equation (or an approximation to it) is solved. The calculation involves both the nuclei and the electrons in the molecule. Quantum mechanics calculations can provide a wealth of information about the properties of the molecule but can be rather time-consuming. Molecular mechanics uses a simpler representation of the molecule. In contrast to quantum mechanics, the energy is dependent solely upon the positions of the nuclei; the electrons are not explicitly represented. In molecular mechanics the energy of a given conformation is calculated using a *force field*. A typical force field will have contributions from the stretching of bonds, from the bending of angles, from torsional rotations and from electrostatic and van der Waals interactions between atoms, as in the following equation [Leach 2001]:

$$V = \sum_{\text{bonds}} \frac{k_i}{2} \left(l_i - l_{i,0}\right)^2 + \sum \frac{k_i}{2} \left(\theta_i - \theta_{i,0}\right)^2 + \sum_{\text{torsions}} \frac{V_n}{2} \left(1 - \cos\left(n\tau - \gamma\right)\right)$$

$$+ \sum_{i=1}^{N} \sum_{j=i+1}^{N} \frac{q_i q_j}{4\pi \varepsilon_0 r_{ij}} + \sum_{i=1}^{N} \sum_{j=i+1}^{N} 4\varepsilon_{ij} \left[\left(\frac{\sigma_{ij}}{r_{ij}}\right)^{12} - \left(\frac{\sigma_{ij}}{r_{ij}}\right)^{6}\right] \quad (A2.1)$$

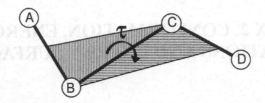

Figure A2-1. The torsion angle A–B–C–D (τ) is defined as the angle between the two planes ABC and BCD. A torsion angle can vary between $0°$ and $360°$, although it is more common to use the range $-180°$ to $+180°$. If one looks along the bond B–C then the torsion angle is the angle through which it is necessary to rotate the bond A–B in a clockwise direction in order to superimpose the two planes.

The various constants in Equation A2.1 (e.g. $k_i, l_{i,0}, V_n, \sigma_{ij}$) are parameters that must be determined, usually from experimental measurements or from high-level quantum mechanics calculations on small representative systems. The first two terms in the equation provide the energy due to the stretching and bending of bonds and variations in bond angles, respectively. These are modelled in Equation A2.1 by a Hooke's law formulation, in which the energy is proportional to the square of the deviation from the relevant reference value (i.e. $l_{i,0}$ or $\theta_{i,0}$). The third term corresponds to changes in torsion angles. An appropriate choice of the parameters V_n, n and γ enables the energy variation due to bond rotations to be reproduced. Perhaps the simplest example is ethane, which shows threefold symmetry with energy minima at $\pm 60°$ and $\pm 180°$ (Figure A2-2).

The fourth and fifth terms in Equation A2.1 provide the non-bonded interactions involving summations over all pairs of atoms (other than those that are bonded to each other or to a common third atom). The fourth term gives the contribution due to electrostatic interactions that arise from the uneven distribution of charge in the molecule. Electrostatic interactions are often modelled using a Coulombic expression that varies as the inverse of the interatomic separation (Figure A2-3). The variables q_i and q_j in the expression are *partial atomic charges*; these are non-integral charge values assigned to each atom in order to reproduce the electrostatic properties of the molecule. The fifth term in Equation A2.1 is that due to van der Waals forces, modelled here using the Lennard–Jones potential. There are two components to these forces. First, there are attractive interactions between the instantaneous dipoles that arise from electronic fluctuations. This is modelled using the $1/r^6$ term. At shorter separations there is a repulsive interaction, quantum mechanical in origin, and modelled using a $1/r^{12}$ term. The combination of these two terms gives rise to the potential curve shown in Figure A2-3.

The variations in energy due to changes in conformation define the molecule's *energy surface* (or *hypersursface*). An energy surface will usually have positions

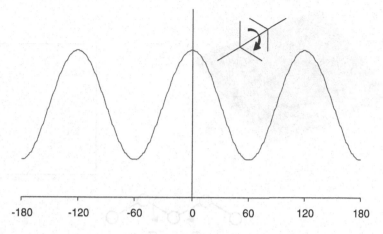

Figure A2-2. The variation in energy with rotation about the carbon–carbon single bond in ethane shows threefold symmetry, with minima at ±60° and ±180° and a barrier height of approximately 2.9 kcal/mol.

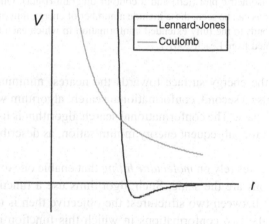

Figure A2-3. Schematic illustration of the van der Waals and electrostatic interactions, modelled using the Lennard–Jones and Coulombic potentials, respectively.

where the energy passes through a minimum; at such a point any change in the conformation will cause the energy to increase. The objective of a conformational search is usually to identify the minimum-energy conformations of a molecule. The minimum with the lowest energy is called the *global minimum-energy conformation*. An example of an energy surface for a simple molecule (*n*-pentane) is shown in Figure A2-4.

A *minimisation algorithm* is a computational procedure that takes as input an initial conformation and changes it to locate the associated minimum point on the energy surface. Most minimisation methods are only able to move

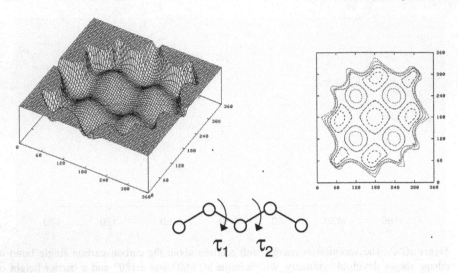

Figure A2-4. Variation in the energy of *n*-pentane with the two torsion angles τ_1 and τ_2, represented as an isometric plot (left) and a contour diagram (right). Only the lowest energy regions are shown; as can be seen, these include a number of energy minima. The global energy minimum corresponds to the fully extended conformation in which each torsion angle has the value 180°. (Adapted from Leach 2001.)

"downhill" on the energy surface towards the nearest minimum. This is why it is necessary to use a second, conformational search, algorithm when exploring the conformational space. The conformational search algorithm is used to generate the initial structures for subsequent energy minimisation, as described in Chapter 2.

Many procedures rely on *molecular fitting* that enable one to calculate whether two conformations are the same. Such algorithms use a function that quantifies the "difference" between two structures; the objective then is to find the relative orientations of the two conformations in which this function is minimised. The most commonly used function is the root mean square distance (RMSD) between pairs of atoms:

$$\text{RMSD} = \sqrt{\frac{\sum_{i=1}^{N_{\text{atoms}}} d_i^2}{N_{\text{atoms}}}} \qquad (A2.2)$$

where N_{atoms} is the number of atoms over which the RMSD is measured and d_i is the distance between atom i in the two structures, when they are overlaid. Some molecular fitting algorithms use an iterative approach, moving one molecule relative to the other in order to reduce the RMSD. Other algorithms locate the optimal position directly.

FURTHER READING

This section lists a few reviews, books and other sources of background material for each chapter.

Chapter 1

Allen F H, G M Battle and S Robertson (2006). The Cambridge Crystallographic Database. In Triggle D J and J B Taylor (Editors), *Comprehensive Medicinal Chemistry II*, Volume 3. Oxford, Elsevier, pp 389–410.

Ash J E, W A Warr and P Willett (Editors) (1991). *Chemical Structure Systems*. Chichester, Ellis Horwood.

Barnard J M (1993). Substructure Searching Methods: Old and New. *Journal of Chemical Information and Computer Sciences* **33**:532–538.

Barnard J M (1998). Structure Representation and Search. In Schleyer P von R, N L Allinger, T Clark, J Gasteiger, P A Kollman, H F Schaefer and P R Shreiner (Editors), *The Encyclopedia of Computational Chemistry*, Volume 4. Chichester, Wiley, pp 2818–2826.

Berks A H, J M Barnard and M P O'Hara (1998). Markush Structure Searching in Patents. In Schleyer P von R, N L Allinger, T Clark, J Gasteiger, P A Kollman, H F Schaefer and P R Shreiner (Editors), *The Encyclopedia of Computational Chemistry*, Volume 3. Chichester, Wiley, pp 1552–1559.

Engel T and E Zass (2006). Chemical Information Systems and Databases. In Triggle D J and J B Taylor (Editors), *Comprehensive Medicinal Chemistry II*, Volume 3. Oxford, Elsevier, pp 265–291.

Miller M A (2002). Chemical Database Techniques in Drug Discovery. *Nature Reviews Drug Discovery* **1**:220–227.

Paris C G (1998). Structure Databases. In Schleyer P von R, N L Allinger, T Clark, J Gasteiger, P A Kollman, H F Schaefer and P R Shreiner (Editors), *The Encyclopedia of Computational Chemistry*, Volume 4. Chichester, Wiley, pp 2771–2785.

Rouvray D H (1991). Origins of Chemical Graph Theory. In Bonchev D and D H Rouvray (Editors), *Chemical Graph Theory*, Amsterdam, Gordon and Breach, pp 1–39.

Warr W A (Editor) (1988). *Chemical Structures. The International Language of Chemistry*. Berlin, Springer.

Zass E (1990). A User's View of Chemical Reaction Information Sources. *Journal of Chemical Information and Computer Sciences* **30**:360–372 (and other papers in this special issue of the journal on Computer Reaction Management in Organic Chemistry).

Chapter 2

Clark D E (Editor) (2000). *Evolutionary Algorithms in Molecular Design*. Weinheim, Wiley-VCH.

Good A C and J S Mason (1995). Three-Dimensional Structure Database Searches. In Lipkowitz K B and D B Boyd (Editors), *Reviews in Computational Chemistry*, Volume 7. New York, VCH, pp 67–117.

Guner O F (Editor) (2000). *Pharmacophore Perception, Development, and Use in Drug Design*. La Jolla, CA, International University Line.

Jones G (1998). Genetic and Evolutionary Algorithms. In Schleyer P von R, N L Allinger, T Clark, J Gasteiger, P A Kollman, H F Schaefer III and P R Schreiner (Editors), *The Encyclopedia of Computational Chemistry*. Chichester, Wiley, pp 1127–1136.

Judson R (1997). Genetic Algorithms and Their Use in Chemistry. In Lipkowitz K B and D B Boyd (Editors), *Reviews in Computational Chemistry*, Volume 10. New York, VCH, pp 1–73.

Leach A R (1991). A Survey of Methods for Searching the Conformational Space of Small and Medium-Sized Molecules. In Lipkowitz K B and D B Boyd (Editors), *Reviews in Computational Chemistry*, Volume 2. New York, VCH, pp 1–55.

Leach A R (2001). *Molecular Modelling Principles and Applications*, Second Edition. Harlow, Pearson Education.

Martin Y C (2006). Pharmacophore Modeling: 1 – Methods. In Triggle D J and J B Taylor (Editors), *Comprehensive Medicinal Chemistry II*, Volume 4. Oxford, Elsevier, pp 119–148.

Martin Y C (2006). Pharmacophore Modeling: 2 – Applications. In Triggle D J and J B Taylor (Editors), *Comprehensive Medicinal Chemistry II*, Volume 4. Oxford: Elsevier, pp 515–536.

Martin Y C, M G Bures and P Willett (1990). Searching Databases of Three-Dimensional Structures. In Lipkowitz K B and D B Boyd (Editors), *Reviews in Computational Chemistry*, Volume 1. New York, VCH, pp 213–263.

Chapter 3

Bath P A, C A Morris and P Willett (1993). Effect of Standardisation on Fragment-Based Measures of Structural Similarity. *Journal of Chemometrics* 7:543–550.

Brown R D (1997). Descriptors for Diversity Analysis. *Perspectives in Drug Discovery and Design* 7/8:31–49.

Downs G M (2003). Molecular Descriptors. In Bultinck P, H De Winter, W Langenaeker and J P Tollenaere (Editors), *Computational Medicinal Chemistry and Drug Discovery*. New York, Marcel Dekker.

Eriksson L, E Johansson, N Kettaneh-Wold and S Wold (1999). *Introduction to Multi- and Megavariate Data Analysis Using Projection Methods*. Umeå, Umetrics.

Hall L H (2006). Topological Quantitative Structure–Activity Relationship Applications: Structure Information Representation in Drug Discovery. In Triggle D J and J B Taylor (Editors), *Comprehensive Medicinal Chemistry II*, Volume 4. Oxford, Elsevier, pp 537–574.

Livingstone D (2000). The Characterization of Chemical Structures Using Molecular Properties. A Survey. *Journal of Chemical Information and Computer Sciences* 40:195–209.

Martin Y C and R S DeWitte (Editors) (1999). Hydrophobicity and Solvation in Drug Design. *Perspectives in Drug Discovery and Design*, Volume 17. Dordrecht, Kluwer Academic.

Martin Y C and R S DeWitte (Editors) (2000). Hydrophobicity and Solvation in Drug Design. *Perspectives in Drug Discovery and Design*, Volume 18. Dordrecht, Kluwer Academic.

Otto M (1998). *Chemometrics: Statistics and Computer Application in Analytical Chemistry*. New York, Wiley-VCH.

Chapter 4

Graham R C (1993). *Data Analysis for the Chemical Sciences. A Guide to Statistical Techniques*. New York, VCH.

Greco G, E Novellino and Y C Martin (1997). Approaches to Three-Dimensional Quantitative Structure–Activity Relationships. In Lipkowitz K B and D B Boyd (Editors), *Reviews in Computational Chemistry*, Volume 11. New York, VCH, pp 183–240.

Hansch C and A Leo (1995). *Exploring QSAR* Washington, DC, American Chemical Society.

Kubinyi H (Editor) (1993). *3D QSAR Drug Design. Theory, Methods and Applications*. Leiden, ESCOM Science.

Kubinyi H (1997). QSAR and 3D QSAR in Drug Design. *Drug Discovery Today* 2:457–467 (Part 1); 538–546 (Part 2).

Kubinyi H (2002). From Narcosis to Hyperspace: The History of QSAR. *Quantitative Structure–Activity Relationships* 21:348–356.

Livingstone D (1995). *Data Analysis for Chemists*. Oxford, Oxford University Press.

Martin Y C (1978). *Quantitative Drug Design. A Critical Introduction*. New York, Marcel Dekker.

Oprea T I and C L Waller (1997). Theoretical and Practical Aspects of Three-Dimensional Quantitative Structure–Activity Relationships. In Lipkowitz K B and D B Boyd (Editors), *Reviews in Computational Chemistry*, Volume 11. New York, VCH, pp 127–182.

Tropsha A (2006). Predictive Quantitative Structure–Activity Relationship Modeling. In Triggle D J and J B Taylor (Editors), *Comprehensive Medicinal Chemistry II*, Volume 4. Oxford, Elsevier, pp 149–166.

Tute M S (1990). History and Objectives of Quantitative Drug Design. In Ramsden C A (Editor), *Comprehensive Medicinal Chemistry*, Volume 4. New York, Pergamon Press, pp 1–31.

Chapter 5

Bajorath J (2001). Selected Concepts and Investigations in Compound Classification, Molecular Descriptor Analysis, and Virtual Screening. *Journal of Chemical Information and Computer Sciences* 41:233–245.

Dean P M (Editor) 1995. *Molecular Similarity in Drug Design*. London, Blackie Academic and Professional.

Downs G M and P Willett (1995). Similarity Searching in Databases of Chemical Structures. In Lipkowitz K B and D B Boyd (Editors), *Reviews in Computational Chemistry*, Volume 7. New York, VCH, pp 1–66.

Good A (2006). Virtual Screening. In Triggle D J and J B Taylor (Editors), *Comprehensive Medicinal Chemistry II*, Volume 4. Oxford, Elsevier, pp 459–494.

Johnson M A and G M Maggiora (1990). *Concepts and Applications of Molecular Similarity*. New York, Wiley.

Lemmen C and T Lengauer (2000). Computational Methods for the Structural Alignment of Molecules. *Journal of Computer-Aided Molecular Design* 14:215–232.

Sheridan R P and S K Kearsley (2002). Why Do We Need So Many Chemical Similarity Search Methods? *Drug Discovery Today* 7:903–911.

Willett P, J M Barnard and G M Downs (1998). Chemical Similarity Searching. *Journal of Chemical Information and Computer Sciences* 38:983–996.

Chapter 6

Dean, P M and R A Lewis (Editors) (1999). *Molecular Diversity in Drug Design*. Dordrecht, Kluwer Academic.

Downs G M and Barnard J M (2002). Clustering Methods and Their Uses in Computational Chemistry. In Lipkowitz K B and D B Boyd (Editors), *Reviews in Computational Chemistry*, Volume 18. New York, VCH, pp 1–40.

Everitt B S (1993). *Cluster Analysis*. London, Wiley.

Gillet V J and P Willett (2001). Dissimilarity-Based Compound Selection for Library Design. In Ghose A K and V N Viswanadhan (Editors), *Combinatorial Library Design and Evaluation for Drug Discovery: Principles, Methods, Software Tools and Applications*. New York, Marcel Dekker, pp 379–398.

Willett P (Editor) (1997). Computational Methods for the Analysis of Molecular Diversity. *Perspectives in Drug Discovery and Design*, Volumes 7/8. Dordrecht, Kluwer.

Willett P (1987). *Similarity and Clustering in Chemical Information Systems*. Letchworth, Research Studies Press.

Chapter 7

Beale R and T Jackson (1990). *Neural Computing: An Introduction*. Bristol, Institute of Physics.

Clark D E (Editor) (2000). *Evolutionary Algorithms in Molecular Design*. Weinheim, Wiley-VCH.

Duda R O, P E Hart and D G Stork (2000). *Pattern Classification*, Second Edition. New York, Wiley.

Gedeck P and P Willett (2001). Visual and Computational Analysis of Structure–Activity Relationships in High-throughput Screening Data. *Current Opinion in Chemical Biology* 5:389–395.

Kohonen T (2001). *Self-Organizing Maps*, Third Edition. Berlin, Springer.

Michalewicz Z and D B Fogel (2000). *How to Solve It: Modern Heuristics*. Germany, Springer.

Peterson K L (2000). Artificial Neural Networks and Their Use in Chemistry. In Lipkowitz K B and D B Boyd (Editors), *Reviews in Computational Chemistry*, Volume 16. New York, VCH, pp 53–140.

Reeves C R (1995). *Modern Heuristic Techniques for Combinatorial Optimisation Problems*. Maidenhead, McGraw-Hill.

Weiss S M and N Indurkhya (1998). *Predictive Data Mining. A Practical Guide*. San Francisco, CA, Morgan Kaufman.

Zupan J and J Gasteiger (1999). *Neural Networks in Chemistry and Drug Design*. Weinheim, Wiley-VCH.

Chapter 8

Böhm H-J and G Schneider (Editors) (2000). *Virtual Screening for Bioactive Molecules*. Weinheim, Wiley-VCH.

Klebe G (Editor) (2000). Virtual Screening: An Alternative or Complement to High Throughput Screening? *Perspectives in Drug Discovery and Design*, Volume 20. Dordrecht, Kluwer.

Muegge I and M Rarey (2001). Small Molecule Docking and Scoring. In Lipkowitz K B and D B Boyd (Editors), *Reviews in Computational Chemistry*, Volume 17. New York, VCH, pp 1–60.

Stouten P F W (2006). Docking and Scoring. In Triggle D J and J B Taylor (Editors), *Comprehensive Medicinal Chemistry II*, Volume 4. Oxford: Elsevier, pp 255–282.

van de Waterbeemd H, D A Smith, K Beaumont and D K Walker (2001). Property-Based Design: Optimisation of Drug Absorption and Pharmacokinetics. *Journal of Medicinal Chemistry* **44**:1313–1333.

van de Waterbeemd H and E Gifford (2003). ADMET in Silico Modelling: Towards Prediction Paradise? *Nature Reviews Drug Discovery* **2**:192–204.

Walters W P, M T Stahl and M A Murcko (1998). Virtual Screening – An Overview. *Drug Discovery Today* **3**:160–178.

Chapter 9

Agrafiotis D K, J C Myslik and F R Salemme (1999). Advances in Diversity Profiling and Combinatorial Series Design. *Molecular Diversity* **4**:1–22.

Drewry D H and S S Young (1999). Approaches to the Design of Combinatorial Libraries. *Chemometrics in Intelligent Laboratory Systems* **48**:1–20.

Ghose A K and V N Viswanadhan (Editors) (2001) *Combinatorial Library Design and Evaluation. Principles, Software Tools and Applications in Drug Discovery*. New York, Marcel Dekker.

Lewis R A, S D Pickett and D E Clark (2000). Computer-Aided Molecular Diversity Analysis and Combinatorial Library Design. In Lipkowitz K B and D B Boyd (Editors), *Reviews in Computational Chemistry*, Volume 16. New York, VCH, pp 1–51.

Martin E J, D C Spellmeyer, R E Critchlow Jr and J M Blaney (1997). Does Combinatorial Chemistry Obviate Computer-Aided Drug Design? In Lipkowitz K B and D B Boyd (Editors), *Reviews in Computational Chemistry*, Volume 10. New York, VCH, pp 75–100.

Mason J S and S D Pickett (2003). Combinatorial Library Design, Molecular Similarity and Diversity Applications. In Abraham D A (Editor), *Burger's Medicinal Chemistry*, Sixth Edition. Chichester, Wiley.

Pickett S D (2006). Library Design: Reactant and Product-Based Approaches. In Triggle D J and J B Taylor (Editors), *Comprehensive Medicinal Chemistry II*, Volume 4. Oxford: Elsevier, pp 337–378.

Schnur D M, A J Tebben and C L Cavallaro (2006). Library Design: Ligand and Structure-Based Principles for Parallel and Combinatorial Libraries. In Triggle D J and J B Taylor (Editors), *Comprehensive Medicinal Chemistry II*, Volume 4. Oxford: Elsevier, pp 307–336.

Spellmeyer D C and P D J Grootenhuis (1999). Recent Developments in Molecular Diversity: Computational Approaches to Combinatorial Chemistry. *Annual Reports in Medicinal Chemistry* **34**:287–296.

REFERENCES

Abagyan R and M Totrov (2001). High-Throughput Docking for Lead Generation. *Current Opinion in Chemical Biology* **5**:375–382.

Abraham M H (1993). Scales of Solute Hydrogen-Bonding: Their Construction and Application to Physicochemical and Biochemical Processes. *Chemical Society Reviews* **22**:73–83.

Abraham M H and J A Platts (2001). Hydrogen Bond Structural Group Constants. *Journal of Organic Chemistry* **66**:3484–3491.

Abraham M H, A Ibrahim, A M Zissmos, Y H Zhao, J Comer and D P Reynolds (2002). Applications of Hydrogen Bonding Calculations in Property Based Drug Design. *Drug Discovery Today* **7**:1056–1063.

ACD. Available Chemicals Directory, MDL Information Systems, Inc. 14600 Catalina Street, San Leandro, CA 94577. Also at http://www.mdli.com.

Adamson G W, J Cowell, M F Lynch, A H W McLure, W G Town and A M Yapp (1973). Strategic Considerations in the Design of a Screening System for Substructure Searches of Chemical Structure Files. *Journal of Chemical Documentation* **13**:153–157.

Agrafiotis D K (1997). Stochastic Algorithms for Maximising Molecular Diversity. *Journal of Chemical Information and Computer Sciences* **37**:841–851.

Agrafiotis D K (2002). Multiobjective Optimisation of Combinatorial Libraries. *Journal of Computer-Aided Molecular Design* **5/6**:335–356.

Agrafiotis D K and V S Lobanov (1999). An Efficient Implementation of Distance-Based Diversity Measures Based on k–d Trees. *Journal of Chemical Information and Computer Sciences* **39**:51–58.

Agrafiotis D K and V S Lobanov (2000). Nonlinear Mapping Networks. *Journal of Chemical Information and Computer Sciences* **40**:1356–1362.

Ahlberg C (1999). Visual Exploration of HTS Databases: Bridging the Gap Between Chemistry and Biology. *Drug Discovery Today* **4**:370–376.

Ajay A, W P Walters and M A Murcko (1998). Can We Learn to Distinguish Between "Drug-like" and "Nondrug-like" Molecules? *Journal of Medicinal Chemistry* **41**:3314–3324.

Ajay A and M A Murcko (1995). Computational Methods to Predict Binding Free Energy in Ligand–Receptor Complexes. *Journal of Medicinal Chemistry* **38**:4951–4967.

Aldenderfer M S and R K Blashfield (1984).*Cluster Analysis*. Beverly Hills CA, Sage University Press.

Allen F H (2002). The Cambridge Structural Database: A Quarter of a Million Crystal Structures and Rising. *Acta Crystallographica* **B58**:380–388. Also at http://www.ccdc.cam.ac.uk.

Allen F H, O Kennard, D G Watson, L Brammer, A G Orpen and R Taylor (1987). Tables of Bond Lengths Determined by X-Ray and Neutron Diffraction. 1. Bond Lengths in Organic Compounds. *Journal of the Chemical Society Perkin Transactions* **II**:S1–S19.

Allen F H, S A Bellard, M D Brice, B A Cartwright, A Doubleday, H Higgs, T Hummelink, B G Hummelink-Peters, O Kennard, W D S Motherwell, J R Rodgers and D G Watson (1979). The Cambridge Crystallographic Data Centre: Computer-Based Search, Retrieval, Analysis and Display of Information. *Acta Crystallographica* **B35**:2331–2339.

Andrea T A and H Kalayeh (1991). Applications of Neural Networks in Quantitative Structure–Activity Relationships of Dihydrofolate-Reductase Inhibitors. *Journal of Medicinal Chemistry* **34**:2824–2836.

A-Razzak M and R C Glen (1992). Applications of Rule-Induction in the Derivation of Quantitative Structure–Activity Relationships. *Journal of Computer-Aided Molecular Design* **6**:349–383.

Attias R (1983). DARC Substructure Search System – a New Approach to Chemical Information. *Journal of Chemical Information and Computer Sciences* **23**:102–108.

Bacha P A, H S Gruver, B K Den Hartog, S Y Tamura and R F Nutt (2002). Rule Extraction from a Mutagenicity Data Set Using Adaptively Grown Phylogenetic-like Trees. *Journal of Chemical Information and Computer Sciences* **42**:1104–1111.

Bajorath J (2002). Integration of Virtual and High-Throughput Screening. *Nature Reviews. Drug Discovery* **1**:882–894.

Bajorath J (2004). *Chemoinformatics: Concepts, Methods, and Tools for Drug Discovery.* Totowa, NJ, Humana Press,

Baldwin J J (1996). *Conference on Libraries and Drug Discovery.* Coronado, CA, January 29–31.

Barker E J, D A Cosgrove, E J Gardiner, V J Gillet, P Kitts and P Willett (2006). Scaffold-Hopping Using Clique Detection Applied to Reduced Graphs. *Journal of Chemical Information and Modeling* **46**:503–511.

Barnard J M (1988). Problems of Substructure Search and Their Solution. In Warr W A (Editor), *Chemical Structures. The International Language of Chemistry.* Berlin, Springer, pp 113–126.

Barnard J M (1993). Substructure Searching Methods: Old and New. *Journal of Chemical Information and Computer Sciences* **33**:532–538.

Barnard J M, G M Downs, A von Scholley-Pfab and R D Brown (2000). Use of Markush Structure Analysis Techniques for Descriptor Generation and Clustering of Large Combinatorial Libraries. *Journal of Molecular Graphics and Modelling* **18**:452–463.

Barnum D, J Greene, A Smellie and P Sprague (1996). Identification of Common Functional Configurations Among Molecules. *Journal of Chemical Information and Computer Sciences* **36**:563–571.

Baroni M, G Costantino, G Cruciani, D Riganelli, R Valigi and S Clementi (1993). Generating Optimal Linear PLS Estimations (GOLPE): An Advanced Chemometric Tool for Handling 3D-QSAR Problems. *Quantitative Structure–Activity Relationships* **12**:9–20.

Basak S C, V R Magnuson, G J Niemi and R R Regal (1988). Determining Structural Similarity of Chemicals Using Graph-Theoretic Indices. *Discrete Applied Mathematics* **19**:17–44.

Bass M B, D F Hopkins, W A N Jaquysh and R L Ornstein (1992). A Method for Determining the Positions of Polar Hydrogens Added to a Protein Structure that Maximizes Protein Hydrogen Bonding. *Proteins: Structure, Function, and Genetics* **12**:266–277.

Baxter C A, C W Murray, D E Clark, D R Westhead and M D Eldridge (1998). Flexible Docking Using Tabu Search and an Empirical Estimate of Binding Affinity. *Proteins: Structure, Function and Genetics* **33**:367–382.

Bayada D M, H Hamersma and V J van Geerestein (1999). Molecular Diversity and Representativity in Chemical Databases. *Journal of Chemical Information and Computer Sciences* **39**:1–10.

Bayley M J and Willett P (1999). Binning Schemes for Partition-Based Compound Selection. *Journal of Molecular Graphics and Modelling* **17**:10–18.

BCI. Barnard Chemical Information Ltd., 46 Uppergate Road, Stannington, Sheffield S6 6BX, UK. Also at http://www.bci.gb.com.

Begg T and C Connely (1998). *Database Systems: An Introduction*, Second Edition. London, Addison-Wesley.

Bemis G W and I D Kuntz (1992). A Fast and Efficient Method for 2D and 3D Molecular Shape Descriptors. *Journal of Computer-Aided Molecular Design* **6**:607–628.

Bender A, H Y Mussa, R C Glen and S Reiling (2004). Molecular Similarity Searching Using Atom Environments: Information-Based Feature Selection and a Naive Bayesian Classifier. *Journal of Chemical Information and Computer Sciences* **44**:170–178.

Benichou P, C Klimczak and P Borne (1997). Handling Genericity in Chemical Structures Using the Markush DARC Software. *Journal of Chemical Information and Computer Sciences* **37**:43–53.

Bennett K P and C Campbell (2000). Support Vector Machines: Hype or Hallelujah? *SIGKDD Explorations* **2**:1–13.

Beresford A P, H E Selick and M H Tarbit (2002). The Emerging Importance of Predictive ADME Simulation in Drug Discovery. *Drug Discovery Today* **7**:109–116.

Berman H M, J Westbrook, Z Feng, G Gilliland, T N Bhat, H Weissig, I N Shindyalov and P E Bourne (2000). The Protein Data Bank. *Nucleic Acids Research* **28**:235–242.

Bernstein F C, T F Koetzle, G J B Williams, E Meyer, M D Bryce, J R Rogers, O Kennard, T Shikanouchi and M Tasumi (1977). The Protein Data Bank: A Computer-Based Archival File for Macromolecular Structures. *Journal of Molecular Biology* **112**:535–542. Also at http://www.rcsb.org.

Beroza P, E K Bradley, J E Eksterowicz, R Feinstein, J Greene, P D J Grootenhuis, R M Henne, J Mount, W A Shirley, A Smellie, R V Stanton and D C Spellmeyer (2000). Applications of Random Sampling to Virtual Screening of Combinatorial Libraries. *Journal of Molecular Graphics and Modelling* **18**:335–342.

Bissantz C, G Folkers and D Rognan (2000). Protein-Based Virtual Screening of Chemical Databases. 1. Evaluations of Different Docking/Scoring Combinations. *Journal of Medicinal Chemistry* **43**:4759–4767.

Blake J E and R C Dana (1990). CASREACT – More than a Million Reactions. *Journal of Chemical Information and Computer Sciences* **30**:394–399.

Blaney F E, C Edge and R W Phippen (1995). Molecular Surface Comparison: 2. Similarity of Electrostatic Surface Vectors in Drug Design. *Journal of Molecular Graphics* **13**:165–174.

Blaney F E, P Finn, R W Phippen and M Wyatt (1993). Molecular Surface Comparison: Application to Molecular Design. *Journal of Molecular Graphics* **11**:98–105.

Blaney J M and J S Dixon (1993). A Good Ligand is Hard to Find: Automated Docking Methods. *Perspectives in Drug Discovery and Design* **1**:301–319.

Blaney J M and J S Dixon (1994). Distance Geometry in Molecular Modeling. In Lipkowitz K B and D B Boyd (Editors), *Reviews in Computational Chemistry*, Volume 5. New York, VCH, pp 299–335.

Böhm H-J (1994). The Development of a Simple Empirical Scoring Function to Estimate the Binding Constant for a Protein–Ligand Complex of Known Three-Dimensional Structure. *Journal of Computer-Aided Molecular Design* **8**:243–256.

Böhm H-J (1998). Prediction of Binding Constants of Protein Ligands: A Fast Method for the Prioritisation of Hits Obtained from De Novo Design or 3D Database Search Programs. *Journal of Computer-Aided Molecular Design* **12**:309–323.

Böhm H-J and G Klebe (1996). What can We Learn from Molecular Recognition in Protein–Ligand Complexes for the Design of New Drugs? *Angewandte Chemie (International ed. in English)* **35**:2588–2614.

Böhm H-J and M Stahl (1999). Rapid Empirical Scoring Functions in Virtual Screening Applications. *Medicinal Chemistry Research* **9**:445–463.

Böhm H-J and M Stahl (2000). Structure-Based Library Design: Molecular Modelling Merges with Combinatorial Chemistry. *Current Opinion in Chemical Biology* 4:283–286.

Böhm H-J, A Flohr and M Stahl (2004). Scaffold Hopping. *Drug Discovery Today: Technologies* 1:217–224.

Borkent J H, F Oukes and J H Noordik (1988). Chemical Reaction Searching Compared in REACCS, SYNLIB, and ORAC. *Journal of Chemical Information and Computer Sciences* 28:148–150.

Boström J (2001). Reproducing the Conformations of Protein-Bound Ligands: A Critical Evaluation of Several Popular Conformational Searching Tools. *Journal of Computer-Aided Molecular Design* 15:1137–1152.

Boström J, P-O Norrby and T Liljefors (1998). Conformational Energy Penalties of Protein-Bound Ligands. *Journal of Computer-Aided Molecular Design* 12:383–396.

Bradshaw J (1997). *Introduction to Tversky*. Also at http://www.daylight.com/meetings/mug97/Bradshaw/MUG97/tv_tversky.html.

Bravi G, D V S Green, M M Hann and A R Leach (2000). Plums: A Program for the Rapid Optimisation of Focussed Libraries. *Journal of Chemical Information and Computer Sciences* 40:1441–1448.

Breiman L (2001). Random Forests. *Machine Learning* 45:5–32.

Breiman L, J H Friedman, R A Olshen and S J Stone (1984). *Classification and Regression Trees*. Belmont, Wadsworth.

Brenner S E, C Chothia and T J P Hubbard (1997). Population Statistics of Protein Structures: Lessons from Structural Classifications. *Current Opinion in Structural Biology* 7:369–376.

Brint A P and P Willett (1987). Algorithms for the Identification of Three-Dimensional Maximal Common Substructures. *Journal of Chemical Information and Computer Sciences* 27:152–158.

Bron C and J Kerbosch (1973). Algorithm 475. Finding All Cliques of an Undirected Graph. *Communications of the ACM* 16:575–577.

Brown N and R A Lewis (2006). Exploiting QSAR Methods in Lead Optimization. *Current Opinion in Drug Discovery and Development* 9:419–424.

Brown F K (1998). Chemoinformatics: What is It and How does It Impact Drug Discovery? *Annual Reports in Medicinal Chemistry* 33:375–384.

Brown J D, M Hassan and M Waldman (2000). Combinatorial Library Design for Diversity, Cost Efficiency, and Drug-like Character. *Journal of Molecular Graphics and Modelling* 18:427–437.

Brown R D and Y C Martin (1996). Use of Structure–Activity Data to Compare Structure-Based Clustering Methods and Descriptors for Use in Compound Selection. *Journal of Chemical Information and Computer Sciences* 36:572–583.

Brown R D and Y C Martin (1997). The Information Content of 2D and 3D Structural Descriptors Relevant to Ligand-Receptor Binding. *Journal of Chemical Information and Computer Sciences* 37:1–9.

Bruccoleri R E and M Karplus (1987). Prediction of the Folding of Short Polypeptide Segments by Uniform Conformational Sampling. *Biopolymers* 26:137–138.

Bruneau P and N R McElroy (2006). logD7.4 Modeling Using Bayesian Regularized Neural Networks. Assessment and Correction of the Errors of Prediction. *Journal of Chemical Information and Computer Sciences* 46:1379–1387.

Brunger A T and M Karplus (1988). Polar Hydrogen Positions in Proteins: Empirical Energy Placement and Neutron Diffraction Comparison. *Proteins: Structure, Function, and Genetics* 4:148–156.

Bruno I J, J C Cole, J P M Lommerse, R S Rowland, R Taylor and M L Verdonk (1997). Isostar: A Library of Information About Nonbonded Interactions. *Journal of Computer-Aided Molecular Design* **11**:525–537.

Burden F R (1989). Molecular Identification Number of Substructure Searches. *Journal of Chemical Information and Computer Sciences* **29**:225–227.

C5.0. RuleQuest Research Pty Ltd. St Ives NSW, Australia. Also at http://www.rulequest.com.

Calinski T and J Harabasz (1974). A Dendrite Method for Cluster Analysis. *Communications in Statistics* **3**:1–27.

Carbó R, L Leyda and M Arnau (1980). An Electron Density Measure of the Similarity Between Two Compounds. *International Journal of Quantum Chemistry* **17**:1185–1189.

Carhart R E, D H Smith and R Venkataraghavan (1985). Atom Pairs as Molecular Features in Structure–Activity Studies: Definition and Applications. *Journal of Chemical Information and Computer Sciences* **25**:64–73.

Carlson H (2002). Protein Flexibility and Drug Design: How to Hit a Moving Target. *Current Opinion in Chemical Biology* **6**:447–452.

CART. Salford Systems, 8880 Rio San Diego Dr., Ste. 1045, San Diego, CA 92108. Also at http://www.salford-systems.com.

Catalyst/HipHop. Accelrys Inc., 9685 Scranton Road, San Diego, CA 92121. Also at http://www.accelrys.com.

CASREACT. Chemical Abstracts Service, 2540 Olentangy River Road, PO Box 3012, Columbus, OH 43210. Also at http://www.cas.org.

Chang G, W C Guida and W C Still (1989). An Internal Coordinate Monte Carlo Method for Searching Conformational Space. *The Journal of the American Chemical Society* **111**:4379–4386.

Charifson P S and W P Walters (2000). Filtering Databases and Chemical Libraries. *Molecular Diversity* **5**:185–197.

Charifson P S, J J Corkery, M A Murcko and W P Walters (1999). Consensus Scoring: A Method for Obtaining Improved Hit Rates from Docking Databases of Three-Dimensional Structures into Proteins. *Journal of Medicinal Chemistry* **42**:5100–5109.

Chau P-L and P M Dean (1987). Molecular Recognition: 3D Surface Structure Comparison by Gnomonic Projection. *Journal of Molecular Graphics* **5**:97–100.

ChemDraw (2002). CambridgeSoft Corporation, 100 Cambridge Park Drive, Cambridge, MA 02140. Also at http://www.cambridgesoft.com.

Chen B N, R F Harrison, K Pasupa, P Willett, D J Wilton, D J Wood and X Q Lewell (2006). Virtual Screening Using Binary Kernel Discrimination: Effect of Noisy Training Data and the Optimization of Performance. *Journal of Chemical Information and Modeling* **46**:478–486.

Chen L, J G Nourse, B D Christie, B A Leland and D L Grier (2002). Over 20 Years of Reaction Access Systems from MDL: A Novel Reaction Substructure Search Algorithm. *Journal of Chemical Information and Computer Sciences* **42**:1296–1310.

Christianini N and J Shawe-Taylor (2000). *An Introduction to Support Vector Machines and Other Kernal-Based Learning Methods*. Cambridge, Cambridge University Press.

Christie B D, B A Leland and J G Nourse (1993). Structure Searching in Chemical Databases by Direct Lookup Methods. *Journal of Chemical Information and Computer Sciences* **33**:545–547.

Clark D E (Editor) (2000). *Evolutionary Algorithms in Molecular Design*. Weinheim, Wiley-VCH.

Clark D E (1999). Rapid Calculation of Polar Surface Area and Its Application to the Prediction of Transport Phenomena. 1. Prediction of Intestinal Absorption. *Journal of Pharmaceutical Sciences* **88**:807–814.

Clark D E and D R Westhead (1996). Evolutionary Algorithms in Computer-Aided Molecular Design. *Journal of Computer-Aided Molecular Design* **10**:337–358.

Clark D E and P D J Grootenhuis (2002). Progress in Computational Methods for the Prediction of ADMET Properties. *Current Opinion in Drug Discovery and Development* **5**:382–390.

Clark D E and S D Pickett (2000). Computational Methods for the Prediction of "Drug-likeness". *Drug Discovery Today* **5**:49–58.

Clark R D (1997). OptiSim: An Extended Dissimilarity Selection Method for Finding Diverse Representative Subsets. *Journal of Chemical Information and Computer Sciences* **37**:1181–1188.

Clark R D and W J Langton (1998). Balancing Representativeness Against Diversity Using Optimisable K-Dissimilarity and Hierarchical Clustering. *Journal of Chemical Information and Computer Sciences* **38**:1079–1086.

Coello C C A, D A van Veldhuizen and G B Lamont (2002). *Evolutionary Algorithms for Solving Multi-Objective Problems*. New York, Kluwer Academic.

Congreve M, R Carr, C Murray and H Jhoti (2003). A "Rule of Three" for Fragment-Based Lead Discovery? *Drug Discovery Today* **8**:876–877.

Cottrell S J, V J Gillet, R Taylor and D J Wilton (2004). Generation of Multiple Pharmacophore Hypotheses Using Multiobjective Optimisation Techniques. *Journal of Computer-Aided Molecular Design* **18**:665–682.

Cox T F and M A A Cox (1994). *Multidimensional Scaling*. London, Chapman & Hall.

Cramer R D III (1993). Partial Least Squares (PLS): Its Strengths and Limitations. *Perspectives in Drug Discovery and Design* **1**:269–278.

Cramer R D III, D E Patterson and J D Bunce (1988). Comparative Molecular Field Analysis (CoMFA). 1. Effect of Shape on Binding of Steroids to Carrier Proteins. *The Journal of the American Chemical Society* **110**:5959–5967.

Cramer R D III, G Redl and C E Berkoff (1974). Substructural Analysis. A Novel Approach to the Problem of Drug Design. *Journal of Medicinal Chemistry* **17**:533–535.

Cringean J K, C A Pepperrell, A R Poirrette and P Willett (1990). Selection of Screens for Three-Dimensional Substructure Searching. *Tetrahedron Computer Methodology* **3**:37–46.

Crippen G M (1981). *Distance Geometry and Conformational Calculations. Chemometrics Research Studies Series 1*. New York, Wiley.

Crippen G M and T F Havel (1988). *Distance Geometry and Molecular Conformation. Chemometrics Research Studies Series 15*. New York, Wiley.

Crivori P, G Cruciani, P-A Carrupt and B Testa (2000). Predicting Blood-Brain Barrier Permeation from Three-Dimensional Molecular Structure. *Journal of Medicinal Chemistry* **43**:2204–2216.

Crossfire. The Crossfire Beilstein Database. MDL Information Systems, Inc. 14600 Catalina Street, San Leandro, CA 94577. Also at http://www.mdli.com.

Cruciani G J, P Crivori, P-A Carrupt and B Testa (2000). Molecular Fields in Quantitative Structure–Permeation Relationships: The VolSurf Approach. *Journal of Molecular Structure: Theochem* **503**:17–30.

Cruciani G, E Carosati, B De Boeck, K Ethirajulu, C Mackie, T Howe and R Vianello (2005). MetaSite: Understanding Metabolism in Human Cytochromes from the Perspective of the Chemist. *Journal of Medicinal Chemistry* **48**: 6970–6979.

Cruciani G, S Clementi and M Baroni (1993). Variable Selection in PLS Analysis. In Kubinyi H (Editor), *3D QSAR* Leiden, ESCOM, pp 551–564.

Cummins D J, C W Andrews, J A Bentley and M Cory (1996). Molecular Diversity in Chemical Databases: Comparison of Medicinal Chemistry Knowledge Bases and Databases of Commercially Available Compounds. *Journal of Chemical Information and Computer Sciences* **36**: 750–763.

Dalby A, J G Nourse, W D Hounshell, A K I Gushurst, D L Grier, B A Leland and J Laufer, (1992). Description of Several Chemical Structure File Formats Used by Computer Programs Developed at Molecular Design Limited. *Journal of Chemical Information and Computer Sciences* **32**:244–255.

Dammkoehler R A, S F Karasek, E F B Shands and G R Marshall (1989). Constrained Search of Conformational Hyperspace. *Journal of Computer-Aided Molecular Design* **3**:3–21.

Danielson E, J H Golden, E W McFarland, C M Reaves, W H Weinberg and X D Wu (1997). A Combinatorial Approach to the Discovery and Optimization of Luminescent Materials. *Nature* **389**:944–948.

Date C J (2000). *An Introduction to Database Systems*. Reading, MA, Addison-Wesley.

Davies E K, M Glick, K N Harrison and W G Richards (2002). Pattern Recognition and Massively Distributed Computing. *Journal of Computational Chemistry* **23**:1544–1550.

Davies K (1996). Using Pharmacophore Diversity to Select Molecules to Test From Commercial Catalogues. In Chaiken I M and K D Janda (Editors), *Molecular Diversity and Combinatorial Chemistry. Libraries and Drug Discovery*. Washington, DC, American Chemical Society, pp 309–316.

Davis L (Editor) (1991). *Handbook of Genetic Algorithms*. New York, Van Nostrand Reinhold.

Daylight. Daylight Chemical Information Systems, Inc., Mission Viejo, CA. Also at http://www.daylight.com.

de Groot M J and S Ekins (2002). Pharmacophore Modeling of Cytochromes P450. *Advanced Drug Delivery Reviews* **54**:367–383.

Desjarlais R L, R P Sheridan, G L Seibel, J S Dixon, I D Kuntz and R Venkataraghavan (1988). Using Shape Complementarity as an Initial Screen in Designing Ligands for a Receptor Binding Site of Known Three-Dimensional Structure. *Journal of Medicinal Chemistry* **31**:722–729.

Dittmar P G, N A Farmer, W Fisanick, R C Haines and J Mockus (1983). The CAS ONLINE Search System. 1. General System Design and Selection, Generation, and Use of Search Screens. *Journal of Chemical Information and Computer Sciences* **23**:93–102.

DISCO. Tripos Inc., 1699 South Hanley Road, St. Louis, MO 63144-2913. Also at http://www.tripos.com.

Dixon S L and R T Koehler (1999). The Hidden Component of Size in Two-Dimensional Fragment Descriptors: Side Effects on Sampling in Bioactive Libraries. *Journal of Medicinal Chemistry* **42**:2887–2900.

Doman T N, S L McGovern, B J Witherbee, T P Kasten, R Kurumbail, W C Stallings, D T Connelly and B K Shoichet (2002). Molecular Docking and High-Throughput Screening for Novel Inhibitors of Protein Tyrosine Phosphatase-1B. *Journal of Medicinal Chemistry* **45**:2213–2221.

Downs G M and J M Barnard (1997). Techniques for Generating Descriptive Fingerprints in Combinatorial Libraries. *Journal of Chemical Information and Computer Sciences* **37**:59–61.

Downs G M and J M Barnard (2002). Clustering Methods and Their Uses in Computational Chemistry. In Lipkowitz K B and D B Boyd (Editors), *Reviews in Computational Chemistry*, Volume 18. New York, VCH, pp 1–40.

Downs G M and P Willett (1994). Clustering of Chemical Structure Databases for Compound Selection. In van de Waterbeemd H (Editor), *Advanced Computer-Assisted Techniques in Drug Discovery*. Weinheim, VCH, pp 111–130.

Downs G M and P Willett (1995). Similarity Searching in Databases of Chemical Structures. In Lipkowitz K B and D B Boyd (Editors), *Reviews in Computational Chemistry*, Volume 7. New York, VCH, pp 1–66.

Downs G M, M F Lynch, P Willett, G Manson and G A Wilson 1988a. Transputer Implementations of Chemical Substructure Searching Algorithms. *Tetrahedron Computer Methodology* **1**:207–217.

Downs G M, P Willett and W Fisanick (1994). Similarity Searching and Clustering of Chemical Structure Databases Using Molecular Property Data. *Journal of Chemical Information and Computer Sciences* **34**:1094–1102.

Downs G M, V J Gillet, J D Holliday and M F Lynch (1989). A Review of Ring Perception Algorithms. *Journal of Chemical Information and Computer Sciences* **29**:172–187.

Downs G M, V J Gillet, J Holliday and M F Lynch (1988b). The Sheffield University Generic Chemical Structures Project – A Review of Progress and of Outstanding Problems. In Warr W A (Editor), *Chemical Structures. The International Language of Chemistry.* Berlin, Springer, pp 151–167.

Duffy E M and W L Jorgensen (2000a). Prediction of Drug Solubility from Monte Carlo Simulations. *Bioorganic and Medicinal Chemistry Letters* **10**:1155–1158.

Duffy E M and W L Jorgensen (2000b). Prediction of Properties from Simulations: Free Energies of Solvation in Hexadecane, Octanol and Water. *The Journal of the American Chemical Society* **122**:2878–2888.

Dunbar J B (1997). Cluster-Based Selection. *Perspectives in Drug Discovery and Design* **7/8**:51–63.

Dunn W J III, S Wold, U Edlund, S Hellberg and J Gasteiger (1984). Multivariate Structure–Activity Relationships Between Data from a Battery of Biological Tests and an Ensemble of Structure Descriptors: The PLS Method. *Quantitative Structure–Activity Relationships* **3**:131–137.

Durant J L, B A Leland, D R Henry and J G Nourse (2002). Reoptimisation of MDL Keys for Use in Drug Discovery. *Journal of Chemical Information and Computer Sciences* **42**:1273–1280.

Durham S K and G M Pearl (2001). Computational Methods to Predict Drug Liabilities. *Current Opinion in Drug Discovery and Development* **4**:110–115.

Ebe T, K A Sanderson and P S Wilson (1991). The Chemical Abstracts Service Generic Chemical (Markush) Structure Storage and Retrieval Capability. 2. The MARPAT File. *Journal of Chemical Information and Computer Sciences* **31**:31–36.

Edgar S, J D Holliday and P Willett (2000). Effectiveness of Retrieval in Similarity Searches of Chemical Databases: A Review of Performance Measures. *Journal of Molecular Graphics and Modelling* **18**:343–357.

Egan W J and G Lauri (2002). Prediction of Intestinal Permeability. *Advanced Drug Delivery Reviews* **54**:273–289.

Ekins S, C L Waller, P W Swaan, G Cruciani, S A Weighton and J H Wikel (2000). Progress in Prediction Human ADME Parameters In Silico. *Journal of Pharmacological and Toxicological Methods* **44**:251–272.

El Tayar N, R-S Tsai, P-A Carrupt and B Testa (1992). Octan-1-ol Water Partition Coefficients of Zwitterionic α-Amino Acids. Determination by Centrifugal Partition Chromatography and Factorisation into Steric/Hydrophobic and Polar Components. *Journal of the Chemical Society Perkin Transactions* **2**:79–84.

Eldridge M D, C W Murray, T R Auton, G V Paoliniand and R P Mee (1997). Empirical Scoring Functions: I. The Development of a Fast Empirical Scoring Function to Estimate the Binding Affinity of Ligands in Receptor Complexes. *Journal of Computer-Aided Molecular Design* **11**:425–445.

Ertl P, B Rohde and P Selzer (2000). Fast Calculation of Molecular Polar Surface Area as a Sum of Fragment-Based Contributions and Its Application to the Prediction of Drug Transport Properties. *Journal of Medicinal Chemistry* **43**:3714–3717.

Everitt B S (1993). *Cluster Analysis*. London, Wiley.

Feng J, A Sanil and S S Young (2006). PharmID: Pharmacophore Identification Using Gibbs Sampling. *Journal of Chemical Information and Modeling* **46**:1352–1359.

Ferguson D M and D J Raber (1989). A New Approach to Probing Conformational Space with Molecular Mechanics: Random Incremental Pulse Search. *The Journal of the American Chemical Society* **111**:4371–4378.

FIRM. Formal Inference-based Recursive Modeling. University of Minnesota, St. Paul, MN. Also at http://www.stat.umn.edu/Research/Software.html.

Fisanick W, A H Lipkus and A Rusinko III (1994). Similarity Searching on CAS Registry Substances. 2. 2D Structural Similarity. *Journal of Chemical Information and Computer Sciences* **34**:130–140.

Fisanick W, K P Cross and A Rusinko (1992). Similarity Searching on CAS Registry Substances. 1. Global Molecular Property and Generic Atom Triangle Geometry Searching. *Journal of Chemical Information and Computer Sciences* **32**:664–674.

Flower D R (1998). On the Properties of Bit String-Based Measures of Chemical Similarity. *Journal of Chemical Information and Computer Sciences* **38**:379–386.

Fonseca C M and P J Fleming (1993). Genetic Algorithms for Multiobjective Optimization: Formulation, Discussion and Generalisation. In S. Forrest (Editor), *Genetic Algorithms: Proceedings of the Fifth International Conference*. San Mateo, CA, Morgan Kaufmann, pp 416–423.

Forgy E (1965). Cluster Analysis of Multivariate Data: Efficiency vs Interpretability of Classifications. *Biometrics* **21**:768–780.

FRED. OpenEye Scientific Software, 3600 Cerrillos Rd., Suite 1107, Santa Fe, NM 87507. Also at http://www.eyesopen.com.

Free S M and J W Wilson (1964). A Mathematical Contribution to Structure–Activity Studies. *Journal of Medicinal Chemistry* **7**:395–399.

Freeland R G, S A Funk, L J O'Korn and G A Wilson (1979). The Chemical Abstracts Service Chemical Registry System. II Augmented Connectivity Molecular Formula. *Journal of Chemical Information and Computer Sciences* **19**:94–98.

Frimurer T M, R Bywater, L Naerum, L N Lauritsen and S Brunak (2000). Improving the Odds in Discriminating "Drug-like" from "Non Drug-like" Compounds. *Journal of Chemical Information and Computer Sciences* **40**:1315–1324.

Fujita T (1990). The Extrathermodynamic Approach to Drug Design. In Ramsden C A (Editor), *Comprehensive Medicinal Chemistry*, Volume 4. New York, Pergamon Press, pp 497–560.

Fujita T, J Iwasa and C. Hansch (1964). A New Substituent Constant, π, Derived from Partition Coefficients. *The Journal of the American Chemical Society* **86**:5175–5180.

GASP. Tripos Inc., 1699 South Hanley Road, St. Louis, MO 63144–2913. Also at http://www.tripos.com.

Gasteiger J and T Engel (Editors) (2003). *Chemoinformatics: A Textbook*. Weinheim, Wiley-VCH.

Gasteiger J, C Rudolph and J Sadowski (1990). Automatic Generation of 3D Atomic Coordinates for Organic Molecules. *Tetrahedron Computer Methodology* **3**:537–547.

Gedeck P and P Willett (2001). Visual and Computational Analysis of Structure–Activity Relationships in High-Throughput Screening Data. *Current Opinion in Chemical Biology* **5**:389–395.

Gehlhaar D K, G M Verkhivker, P A Rejto, C J Sherman, D B Fogel, L J Fogel and S T Freer (1995). Molecular Recognition of the Inhibitor AG-1343 by HIV-1 Protease: Conformationally Flexible Docking by Evolutionary Programming. *Chemistry and Biology* **2**:317–324.

Ghose A K and G M Crippen (1986). Atomic Physicochemical Parameters for Three-Dimensional Structure-Directed Quantitative Structure–Activity Relationships. I. Partition Coefficients as a Measure of Hydrophobicity. *Journal of Computational Chemistry* **7**:565–577.

Ghose A K, V N Viswanadhan and J J Wendoloski (1998). Prediction of Hydrophobic (Lipophilic) Properties of Small Organic Molecules Using Fragmental Methods: An Analysis of ALOGP and CLOGP Methods. *Journal of Physical Chemistry* **102**:3762–3772.

Gibson K D and H A Scheraga (1987). Revised Algorithms for the Build-Up Procedure for Predicting Protein Conformations by Energy Minimization. *Journal of Computational Chemistry* **8**:826–834.

Gillet V J, G M Downs, J D Holliday, M F Lynch and W Dethlefsen (1991). Computer Storage and Retrieval of Generic Chemical Structures in Patents. 13. Reduced Graph Generation. *Journal of Chemical Information and Computer Sciences* **31**:260–270.

Gillet V J, and A R Leach (2006). Chemoinformatics. In Triggle D J and J B Taylor (Editors), Comprehensive Medicinal Chemistry II, Vol. 3, Oxford, Elsevier, pp. 235–264

Gillet V J, P Willett and J Bradshaw (1997). The Effectiveness of Reactant Pools for Generating Structurally Diverse Combinatorial Libraries. *Journal of Chemical Information and Computer Sciences* **37**:731–740.

Gillet V J, P Willett and J Bradshaw (1998). Identification of Biological Activity Profiles Using Substructural Analysis and Genetic Algorithms. *Journal of Chemical Information and Computer Sciences* **38**:165–179.

Gillet V J, P Willett and J Bradshaw (2003). Similarity Searching Using Reduced Graphs. *Journal of Chemical Information and Computer Sciences* **43**:338–345.

Gillet V J, P Willett, J Bradshaw and D V S Green (1999). Selecting Combinatorial Libraries to Optimize Diversity and Physical Properties. *Journal of Chemical Information and Computer Sciences* **39**:169–177.

Gillet V J, P Willett, P Fleming and D V S Green 2002a. Designing Focused Libraries Using MoSELECT. *Journal of Molecular Graphics and Modelling* **20**:491–498.

Gillet V J, W Khatib, P Willett, P Fleming and D V S Green 2002b. Combinatorial Library Design Using a Multiobjective Genetic Algorithm. *Journal of Chemical Information and Computer Sciences* **42**:375–385.

Ginn C M R, P Willett and J Bradshaw (2000). Combination of Molecular Similarity Measures Using Data Fusion. *Perspectives in Drug Discovery and Design* **20**:1–16.

Glick M, J L Jenkins, J H Nettles, H Hitchings and J W Davies (2006). Enrichment of High-Throughput Screening Data with Increasing Levels of Noise Using Support Vector Machines, Recursive Partitioning, and Laplacian-Modified Naive Bayesian Classifiers. *Journal of Chemical Information and Modeling* **46**:193–200.

Glide. Schrödinger, 1500 S.W. First Avenue, Suite 1180, Portland, OR 97201. Also at http://www.schrodinger.com.

Gohlke H and G Klebe (2001). Statistical Potentials and Scoring Functions Applied to Protein–Ligand Binding. *Current Opinion in Structural Biology* **11**:231–235.

Gohlke H, M Hendlich and G Klebe (2000). Knowledge-Based Scoring Function to Predict Protein–Ligand Interactions. *Journal of Molecular Biology* **295**:337–356.

Golbraikh A and A Tropsha (2002). Beware of q^2! *Journal of Molecular Graphics and Modelling* **20**:269–276.

Goldberg D E (1989). *Genetic Algorithms in Search, Optimization and Machine Learning*. Wokingham, Addison-Wesley.

Gombar V K and K Enslein (1995). Use of Predictive Toxicology in the Design of New Chemicals. *ACS Symposium Series 589 (Computer-Aided Molecular Design)*. Washington, DC, American Chemical Society, pp 236–349.

Good A and R A Lewis (1997). New Methodology for Profiling Combinatorial Libraries and Screening Sets: Cleaning up the Design with HARPick. *Journal of Medicinal Chemistry* **40**:3926–3936.

Good A C and W G Richards (1998). Explicit Calculation of 3D Molecular Similarity. *Perspectives in Drug Discovery and Design* **9/10/11**:321–338.

Good A C, E E Hodgkin and W G Richards (1992). The Utilisation of Gaussian Functions for the Rapid Evaluation of Molecular Similarity. *Journal of Chemical Information and Computer Sciences* **32**:188–192.

Good A C, J S Mason, D V S Green and A R Leach (2001). Pharmacophore-Based Approaches to Combinatorial Library Design. In Ghose A K and V N Viswanadhan (Editors), *Combinatorial Library Design and Evaluation. Principles, Software Tools and Applications in Drug Discovery*. New York, Marcel Dekker, pp 399–428.

Goodford P J (1985). A Computational Procedure for Determining Energetically Favorable Binding Sites on Biologically Important Macromolecules. *Journal of Medicinal Chemistry* **28**:849–857.

Goodman J M and W C Still (1991). An Unbounded Systematic Search of Conformational Space. *Journal of Computational Chemistry* **12**:1110–1117.

Goodsell D S and A J Olson (1990). Automated Docking of Substrates to Proteins by Simulated Annealing. *Proteins: Structure, Function and Genetics* **8**:195–202.

Grant J A and B T Pickup (1995). A Gaussian Description of Molecular Shape. *Journal of Physical Chemistry* **99**:3503–3510.

Grant J A, M A Gallardo and B T Pickup (1996). A Fast Method of Molecular Shape Comparison: A Simple Application of a Gaussian Description of Molecular Shape. *Journal of Computational Chemistry* **17**:1653–1666.

Greene J, S Kahn, H Savoj, P Sprague and S Teig (1994). Chemical Function Queries for 3D Database Search. *Journal of Chemical Information and Computer Sciences* **34**:1297–1308.

Greene N (2002). Computer Systems for the Prediction of Toxicity: An Update. *Advanced Drug Delivery Reviews* **54**:417–431.

Guenoche A, P Hansen and B Jaumard (1991). Efficient Algorithms for Divisive Hierarchical Clustering with the Diameter Criterion. *Journal of Classification* **8**:5–30.

Ha S, R Andreani, A Robbins and I Muegge (2000). Evaluation of Docking/Scoring Approaches: A Comparative Study Based on MMP3 Inhibitors. *Journal of Computer Aided Molecular Design* **14**:435–448.

Hagadone T R (1992). Molecular Substructure Similarity Searching: Efficient Retrieval in Two-Dimensional Structure Databases. *Journal of Chemical Information and Computer Sciences* **32**:515–521.

Hahn M (1997). Three-Dimensional Shape-Based Searching of Conformationally Flexible Compounds. *Journal of Chemical Information and Computer Sciences* **37**:80–86.

Hall L H and L B Kier (1991). The Molecular Connectivity Chi Indexes and Kappa Shape Indexes in Structure-Property Modeling. In Lipkowitz K B and D B Boyd (Editors), *Reviews in Computational Chemistry*, Volume 2. New York, VCH, pp 367–422.

Hall L H, B Mohney and L B Kier (1991). The Electrotopological State: An Atom Index for QSAR. *Quantitative Structure–Activity Relationships* **10**:43–51.

Halperin I, B Ma, H Wolfson and R Nussinov (2002). Principles of Docking: An Overview of Search Algorithms and a Guide to Scoring Functions. *Proteins: Structure, Function and Genetics* **47**:409–443.

Hammett L P (1970). *Physical Organic Chemistry: Reaction Rates, Equilibria, and Mechanisms*. New York, McGraw-Hill.

Hann M M, A R Leach and G Harper (2001). Molecular Complexity and Its Impact on the Probability of Finding Leads for Drug Discovery. *Journal of Chemical Information and Computer Sciences* **41**:856–864.

Hann M, B Hudson, X Lewell, R Lifely, L Miller and N Ramsden (1999). Strategic Pooling of Compounds for High-Throughput Screening. *Journal of Chemical Information and Computer Sciences* **39**:897–902.

Hann, M and R Green (1999). Chemoinformatics – A New Name for an Old Problem? *Current Opinion in Chemistry and Biology* **3**:379–383.

Hansch C (1969). A Quantitative Approach to Biochemical Structure–Activity Relationships. *Accounts of Chemical Research* **2**:232–239.

Hansch C and A Leo (1995). *Exploring QSAR*. Washington, DC, American Chemical Society.

Hansch C and T E Klein (1986). Molecular Graphics and QSAR in the Study of Enzyme–Ligand Interactions. On the Definition of Bioreceptors. *Accounts of Chemical Research* **19**:392–400.

Hansch C and T Fujita (1964). $\rho - \sigma - \pi$ Analysis – A Method for the Correlation of Biological Activity and Chemical Structure. *The Journal of the American Chemical Society* **86**:1616–1626.

Hansch C, A S Ruth, S M Anderson and D L Bentley (1968). Parabolic Dependence of Drug Action upon Lipophilic Character as Revealed by a Study of Hypnotics. *Journal of Medicinal Chemistry* **11**:1–11.

Hansch C, J McClarin, T Klein and R Langridge (1985). A Quantitative Structure–Activity Relationship and Molecular Graphics Study of Carbonic Anhydrase Inhibitors. *Molecular Pharmacology* **27**:493–498.

Hansch C, T Klein, J McClarin, R Langridge and N W Cornell (1986). A Quantitative Structure–Activity Relationship and Molecular Graphics Analysis of Hydrophobic Effects in the Interactions of Inhibitors of Alcohol Dehydrogenase. *Journal of Medicinal Chemistry* **29**:615–620.

Harper G, G S Bravi, S D Pickett, J Hussain and D V S Green (2004). The Reduced Graph Descriptor in Virtual Screening and Data-Driven Clustering of High-Throughput Screening Data. *Journal of Chemical Information and Computer Sciences* **44**:2145–2156.

Harper G, J Bradshaw, J C Gittins, D V S Green and A R Leach (2001). Prediction of Biological Activity for High-Throughput Screening Using Binary Kernel Discrimination. *Journal of Chemical Information and Computer Sciences* **41**:1295–1300.

Harper G, S D Pickett and D V S Green (2004). Design of a Compound Screening Collection for Use in High Throughput Screening. *Combinatorial Chemistry and High Throughput Screening* **7**:63–70.

Hassan M, J P Bielawski, J C Hempel and M Waldman (1996). Optimisation and Visualisation of Molecular Diversity of Combinatorial Libraries. *Molecular Diversity* **2**:64–74.

Hawkins D M (1997). *FIRM*. Technical Report 546, School of Statistics, University of Minnesota, St. Paul, MN.

Hawkins D M, S S Young and A Rusinko III (1997). Analysis of a Large Structure-Activity Data Set Using Recursive Partitioning. *Quantitative Structure–Activity Relationships* **16**:296–302.

Head R D, M L Smythe, T I Oprea, C L Waller, S M Green and G R Marshall (1996). VALIDATE: A New Method for the Receptor-Based Prediction of Binding Affinities of Novel Ligands. *The Journal of the American Chemical Society* **118**:3959–3969.

Hendlich M (1998). Databases for Protein–Ligand Complexes. *Acta Crystallographica, Section D: Biological Crystallography* **D54**:1178–1182.

Hendrickson J B and L Zhang (2000). Duplications Among Reaction Databases. *Journal of Chemical Information and Computer Sciences* **40**:380–383.

Hert J, P Willett, D J Wilton, P Acklin, K Azzaoui, E Jacoby and A Schuffenhauer (2004a). Comparison of Fingerprint-Based Methods for Virtual Screening Using Multiple Bioactive Reference Structures. *Journal of Chemical Information and Computer Sciences* **44**:1177–1185.

Hert J, P Willett, D J Wilton, P Acklin, K Azzaoui, E Jacoby and A Schuffenhauer (2004b). Comparison of Topological Descriptors for Similarity-Based Virtual Screening Using Multiple Bioactive Reference Structures. *Organic and Biomolecular Chemistry* **2**:3256–3266.

Hert J, P Willett, D J Wilton, P Acklin, K Azzaoui, E Jacoby and A Schuffenhauer (2005). Enhancing the Effectiveness of Similarity-Based Virtual Screening Using Nearest-Neighbour Information. *Journal of Medicinal Chemistry* **48**:7049–7054.

Hert J, P Willett, D J Wilton, P Acklin, K Azzaoui, E Jacoby and A Schuffenhauer (2006). New Methods for Ligand-Based Virtual Screening: Use of Data Fusion and Machine Learning to Enhance the Effectiveness of Similarity Searching. *Journal of Chemical Information and Modeling* **46**:462–470.

Hertzberg R P and A J Pope (2000). High-Throughput Screening: New Technology for the 21st Century. *Current Opinion in Chemical Biology* **4**:445–451.

Hicks M G and C Jochum (1990). Substructure Search Systems. 1. Performance Comparison of the MACCS, DARC, HTSS, CAS Registry MVSSS and S4 Substructure Search Systems. *Journal of Chemical Information and Computer Sciences* **30**:191–199.

Hodes L. (1976). Selection of Descriptors According to Discrimination and Redundancy. Application to Chemical Structure Searching, *Journal of Chemical Information and Computer Sciences* **16**:88–93.

Hodes L, G F Hazard, R I Geran and S Richman (1977). Statistical-Heuristic Method for Automated Selection of Drugs for Screening. *Journal of Medicinal Chemistry* **20**:469–475.

Hodgkin E E and W G Richards (1987). Molecular Similarity Based on Electrostatic Potential and Electric Field. *International Journal of Quantum Chemistry. Quantum Biology Symposia* **14**:105–110.

Hodgson J (2001). ADMET – Turning Chemicals into Drugs. *Nature Biotechnology* **19**:722–726.

Holliday J D and M F Lynch (1995). Computer Storage and Retrieval of Generic Chemical Structures in Patents. 16. The Refined Search: An Algorithm for Matching Components of Generic Chemical Structures at the Atom-Bond Level. *Journal of Chemical Information and Computer Sciences* **35**:1–7.

Holliday J D, S R Ranade and P Willett (1995). A Fast Algorithm for Selecting Sets of Dissimilar Molecules from Large Chemical Databases. *Quantitative Structure–Activity Relationships* **14**:501–506.

Holliday J D, V J Gillet, G M Downs, M F Lynch and W Dethlefsen (1992). Computer Storage and Retrieval of Generic Chemical Structures in Patents. 14. Algorithmic Generation of Fragment Descriptors for Generic Structures. *Journal of Chemical Information and Computer Sciences* **32**:453–462.

Holm L and C Sander (1994). The FSSP Database of Structurally Aligned Protein Fold Families. *Nucleic Acids Research* **22**:3600–3609.

Hopkins A L, C R Groom and A Alex (2004). Ligand Efficiency: A Useful Metric for Lead Selection. *Drug Discovery Today* **9**:430–431.

Hudson B D, R M Hyde, E Rahr, J Wood and J Osman (1996). Parameter Based Methods for Compound Selection from Chemical Databases. *Quantitative Structure–Activity Relationships* **15**:285–289.

Hurst T. (1994). The Directed Tweak Technique. *Journal of Chemical Information and Computer Sciences* **34**:190–196.

InChI. The IUPAC Chemical Identifier Project. Also at http://www.iupac.org/inchi/ and http://www.iupac.org/projects/2000/2000–025–1–800.html.

ISIS/Draw (2002). MDL Information Systems, Inc. 14600 Catalina Street, San Leandro, CA 94577. Also at http://www.mdli.com.

James C A, D Weininger and J Delany (2002). *Daylight Theory Manual.* Also at http://www.daylight.com/dayhtml/doc/theory/theory.toc.html.

Jamois E A, M Hassan and M Waldman (2000). Evaluation of Reagent-Based and Product-Based Strategies in the Design of Combinatorial Library Subsets. *Journal of Chemical Information and Computer Sciences* **40**:63–70.

Jarvis R A and E A Patrick (1973). Clustering Using a Similarity Measure Based on Shared Near Neighbours. *IEEE Transactions in Computers* **C-22**:1025–1034.

Jenkins J L, M Glick and J W Davies (2004). A 3D Similarity Method for Scaffold Hopping from Known Drugs or Natural Ligands to New Chemotypes. *Journal of Medicinal Chemistry* **47**:6144–6159.

Johnson M A and G M Maggiora (1990). *Concepts and Applications of Molecular Similarity.* New York, Wiley.

Jones G (1998). Genetic and Evolutionary Algorithms. In Schleyer P von R, N L Allinger, T Clark, J Gasteiger, P A Kollman, H F Schaefer III and P R Schreiner (Editors), *The Encyclopedia of Computational Chemistry.* Chichester, Wiley, pp 1127–1136.

Jones G, P Willett and R C Glen (1995a). A Genetic Algorithm for Flexible Molecular Overlay and Pharmacophore Elucidation. *Journal of Computer-Aided Molecular Design* **9**:532–549.

Jones G, P Willett and R C Glen (1995b). Molecular Recognition of Receptor-Sites Using a Genetic Algorithm with a Description of Desolvation. *Journal of Molecular Biology* **245**:43–53.

Jones G, P Willett and R C Glen (2000). GASP: Genetic Algorithm Superposition Program. In Guner O F (Editor), *Pharmacophore Perception, Development, and Use in Drug Design.* La Jolla, CA, International University Line, pp 85–106.

Jones G, P Willett, R C Glen, A R Leach and R Taylor (1997). Development and Validation of a Genetic Algorithm for Flexible Docking. *Journal of Molecular Biology* **267**:727–748.

Jorgensen W L and E M Duffy (2002). Prediction of Drug Solubility from Structure. *Advanced Drug Delivery Reviews* **54**:355–366.

Judson R S, E P Jaeger and A M Treasurywala (1994). A Genetic Algorithm-Based Method for Docking Flexible Molecules. *Journal of Molecular Structure: Theochem* **114**:191–206.

Kearsley S K and G M Smith (1990). An Alternative Method for the Alignment of Molecular Structures: Maximizing Electrostatic and Steric Overlap. *Tetrahedron Computer Methodology* **3**:615–633.

Kearsley, S K, S Sallamack, E Fluder, J D Andose, R T Mosley and R P Sheridan (1996). Chemical Similarity Using Physiochemical Property Descriptors. *Journal of Chemical Information and Computer Sciences* **36**:118–127.

Kelley L A, S P Gardner and M J Sutcliffe (1996). An Automated Approach for Clustering an Ensemble of NMR-Derived Protein Structures into Conformationally-Related Subfamilies. *Protein Engineering* **9**:1063–1065.

Kick E K, D C Roe, A G Skillman, G C Liu, T J A Ewing, Y X Sun, I D Kuntz and J A Ellman (1997). Structure-Based Design and Combinatorial Chemistry Yield Low Nanomolar Inhibitors of Cathepsin D. *Chemistry and Biology* **4**:297–307.

Kier L B and L H Hall (1986). *Molecular Connectivity in Structure–Activity Analysis.* New York, Wiley.

Kirkpatrick S, C D Gelatt and M P Vecchi (1983). Optimization by Simulated Annealing. *Science* **220**:671–680.

Klebe G, U Abraham and T Mietzner (1994). Molecular Similarity Indices in a Comparative Analysis (CoMSIA) of Drug Molecules to Correlate and Predict their Biological Activity. *Journal of Medicinal Chemistry* **37**:4130–4146.

Klopman G (1984). Artificial Intelligence Approach to Structure-Activity Studies. Computer Automated Structure Evaluation of Biological Activity of Organic Molecules. *The Journal of the American Chemical Society* **106**:7315–7321.

Klopman G (1992). MULTICASE 1. A Hierarchical Computer Automated Structure Evaluation Program. *Quantitative Structure–Activity Relationships* **11**:176–184.

Klopman G and H Zhu (2001). Estimation of the Aqueous Solubility of Organic Molecules by the Group Contribution Approach. *Journal of Chemical Information and Computer Sciences* **41**:439–445.

Klopman G and M Tu (1999). Diversity Analysis of 14156 Molecules Testing by the National Cancer Institute for Anti-HIV Activity Using the Quantitative Structure–Activity Relational Expert System MCASE. *Journal of Medicinal Chemistry* **42**:992–998.

Klopman G, S Wang and D M Balthasar (1992). Estimation of Aqueous Solubility of Organic Molecules by the Group Contribution Approach. Application to the Study of Biodegradation. *Journal of Chemical Information and Computer Sciences* **32**:474–482.

Kontoyianni M, L McClellan and G S Sokol (2004). Evaluation of Docking Performance: Comparative Data on Docking Algorithms. *Journal of Medicinal Chemistry* **47**:558–565.

Kramer B, M Rarey and T Lengauer (1999). Evaluation of the FlexX Incremental Construction Algorithm for Protein–Ligand Docking. *Proteins: Structure, Function and Genetics* **37**:228–241.

Kruskal J B (1964). Multidimensional Scaling by Optimizing Goodness of Fit to a Nonmetric Hypothesis. *Psychometrika* **29**:1–27.

Kubinyi H (1976). Quantitative Structure–Activity Relations. IV. Non-Linear Dependence of Biological Activity on Hydrophobic Character: A New Model. *Arzneimittel-Forschung* **26**:1991–1997.

Kubinyi H (1977). Quantitative Structure–Activity Relations. 7. The Bilinear Model, A New Model for Nonlinear Dependence of Biological Activity on Hydrophobic Character. *Journal of Medicinal Chemistry* **20**:625–629.

Kubinyi H (1998a). Comparative Molecular Field Analysis (CoMFA). In Schleyer P von R, N L Allinger, T Clark, J Gasteiger, P A Kollman, H F Schaefer III and P R Schreiner (Editors), *The Encyclopedia of Computational Chemistry*. Chichester, Wiley, pp 448–460.

Kubinyi H (1998b). Combinatorial and Computational Approaches in Structure-Based Drug Design. *Current Opinion in Drug Discovery and Development* **1**:16–27.

Kuntz I D (1992). Structure-Based Strategies for Drug Design and Discovery. *Science* **257**:1078–1082.

Kuntz I D, E C Meng and B K Shoichet (1994). Structure-Based Molecular Design. *Accounts of Chemical Research* **27**:117–123.

Kuntz I D, J M Blaney, S J Oatley, R Langridge and T E Ferrin (1982). A Geometric Approach to Macromolecule–Ligand Interactions. *Journal of Molecular Biology* **161**:269–288.

Lajiness M (1991). Evaluation of the Performance of Dissimilarity Selection Methodology. In Silipo C and A Vittoria (Editors), *QSAR* Amsterdam, Elsevier Science, pp 201–204.

Lajiness M S (1990). Molecular Similarity-Based Methods for Selecting Compounds for Screening. In Rouvray D H (Editor), *Computational Chemical Graph Theory*. New York, Nova Science, pp 299–316.

Lance G N and W T Williams (1967). A General Theory of Classificatory Sorting Strategies. 1. Hierarchical Systems. *Computer Journal* **9**:373–380.

Leach A R (1991). A Survey of Methods for Searching the Conformational Space of Small and Medium-Sized Molecules. In Lipkowitz K B and D B Boyd (Editors), *Reviews in Computational Chemistry*, Volume 2. New York, VCH, pp 1–55.

Leach A R (2001). *Molecular Modelling Principles and Applications*, Second Edition. Harlow, Pearson Education.

Leach A R and I D Kuntz (1990). Conformational Analysis of Flexible Ligands in Macromolecular Receptor Sites. *Journal of Computational Chemistry* 13:730–748.

Leach A R, B K Shoichet and C E Peishoff (2006). Prediction of Protein–Ligand Interactions. Docking and Scoring: Successes and Gaps. *Journal of Medicinal Chemistry* 49:5851–5855.

Leach A R, D P Dolata and K Prout (1987). WIZARD: AI in Conformational Analysis. *Journal of Computer-Aided Molecular Design* 1:73–85.

Leach A R, J Bradshaw, D V S Green, M M Hann and J J Delany III (1999). Implementation of a System for Reagent Selection and Library Enumeration, Profiling and Design. *Journal of Chemical Information and Computer Sciences* 39:1161–1172.

Leach A R, K Prout and D P Dolata. (1988). An Investigation into the Construction of Molecular Models Using the Template Joining Method. *Journal of Computer-Aided Molecular Design* 2:107–123.

Lebl M (1999). Parallel Personal Comments on Classical Papers in Combinatorial Chemistry. *Journal of Combinatorial Chemistry* 1:3–24.

Lemmen C and T Lengauer (2000). Computational Methods for the Structural Alignment of Molecules. *Journal of Computer-Aided Molecular Design* 14:215–232.

Leo A and Weininger A (1995). *CMR3 Reference Manual*. Also at http:// www.daylight. com/dayhtml/doc/cmr/cmrref.html.

Leo A J (1993). Calculating log *P*oct from Structures. *Chemical Reviews* 93:1281–1306.

Leo A J and D Hoekman (2000). Calculating log *P*(oct) with No Missing Fragments; The Problem of Estimating New Interaction Parameters. *Perspectives in Drug Discovery and Design* 18:19–38.

Lewis R A, J S Mason and I M McLay (1997). Similarity Measures for Rational Set Selection and Analysis of Combinatorial Libraries: The Diverse Property-Derived (DPD) Approach. *Journal of Chemical Information and Computer Sciences* 37:599–614.

Li Z Q and H A Scheraga (1987). Monte-Carlo-Minimization Approach to the Multiple-Minima Problem in Protein Folding. *Proceedings of the National Academy of Sciences of the United States of America* 84:6611–6615.

Lipinski C A (2000). Drug-like Properties and the Causes of Poor Solubility and Poor Permeability. *Journal of Pharmacological and Toxicological Methods* 44:235–249.

Lipinski C A, F Lombardo, B W Dominy and P J Feeney (1997). Experimental and Computational Approaches to Estimate Solubility and Permeability in Drug Discovery and Development Settings. *Advanced Drug Delivery Reviews* 23:3–25.

Lipton M and W C Still (1988). The Multiple Minimum Problem in Molecular Modeling. Tree Searching Internal Coordinate Conformational Space. *Journal of Computational Chemistry* 9:343–355.

Livingstone D (1995), *Data Analysis for Chemists*, Oxford, Oxford University Press

Livingstone D J (2000). The Characterization of Chemical Structures Using Molecular Properties. A Survey. *Journal of Chemical Information and Computer Sciences* 40:195–209.

Livingstone D J and E Rahr (1989). CORCHOP – An Interactive Routine for the Dimension Reduction of Large QSAR Data Sets. *Quantitative Structure–Activity Relationships* 8:103–108.

Lowell H, L H Hall and L B Kier (2001). Issues in Representation of Molecular Structure: The Development of Molecular Connectivity. *Journal of Molecular Graphics and Modelling* 20:4–18.

Luckenbach R (1981). The Beilstein Handbook of Organic Chemistry: The First Hundred Years. *Journal of Chemical Information and Computer Sciences* **21**:82.

Lynch M F and P Willett (1978). The Automatic Detection of Chemical Reaction Sites. *Journal of Chemical Information and Computer Sciences* **18**:154–159.

Lynch M F and J D Holliday (1996). The Sheffield Generic Structures Project – A Retrospective Review. *Journal of Chemical Information and Computer Sciences* **36**:930–936.

Lyne P D (2002). Structure-Based Virtual Screening: An Overview. *Drug Discovery Today* **7**:1047–1055.

Manallack D T, D D Ellis and D J Livingstone (1994). Analysis of Linear and Nonlinear QSAR Data Using Neural Networks. *Journal of Medicinal Chemistry* **37**:3758–3767.

Mannhold R and H van der Waterbeemd (2001). Substructure and Whole Molecule Approaches for Calculating log *P*. *Journal of Computer-Aided Molecular Design* **15**:337–354.

Marriott D P, I G Dougall, P Meghani, Y-J Liu and D R Flower (1999). Lead Generation Using Pharmacophore Mapping and Three-Dimensional Database Searching: Application to Muscarinic M₃ Receptor Antagonists. *Journal of Medicinal Chemistry* **42**:3210–3216.

Martin E J, J M Blaney, M A Siani, D C Spellmeyer, A K Wong and W H Moos (1995). Measuring Diversity: Experimental Design of Combinatorial Libraries for Drug Discovery. *Journal of Medicinal Chemistry* **38**:1431–1436.

Martin Y C (2000). DISCO:What We Did Right and What We Missed. In Guner O F (Editor), *Pharmacophore Perception, Development, and Use in Drug Design*. La Jolla, CA, International University Line, pp 51–66.

Martin Y C and R S DeWitte (Editors) (1999). Hydrophobicity and Solvation in Drug Design. *Perspectives in Drug Discovery and Design*, Volume 17. Dordrecht, Kluwer Academic.

Martin Y C and R S DeWitte (Editors) (2000). Hydrophobicity and Solvation in Drug Design. *Perspectives in Drug Discovery and Design*, Volume 18. Dordrecht, Kluwer Academic.

Martin Y C, J B Holland, C H Jarboe and N Plotnikoff (1974). Discriminant Analysis of the Relationship Between Physical Properties and the Inhibition of Monoamine Oxidase by Aminotetralins and Aminoindans. *Journal of Medicinal Chemistry* **17**:409–413.

Martin Y C, J L Kofron and L M Traphagen (2002). Do Structurally Similar Molecules have Similar Biological Activity? *Journal of Medicinal Chemistry* **45**:4350–4358.

Martin Y C, M G Bures, A A Danaher, J DeLazzer, I Lico and P A Pavlik (1993). A Fast New Approach to Pharmacophore Mapping and its Application to Dopaminergic and Benzodiazepine Agonists. *Journal of Computer-Aided Molecular Design* **7**:83–102.

Mason J S, I M McLay and R A Lewis (1994). Applications of Computer-Aided Drug Design Techniques to Lead Generation. In Dean D M, G Jolles and C G Newton (Editors), *New Perspectives in Drug Design*. London, Academic Press, pp 225–253.

Mason J S, I Morize, P R Menard, D L Cheney, C Hulme and R F Labaudiniere (1999). New 4-Point Pharmacophore Method for Molecular Similarity and Diversity Applications: Overview of the Method and Applications, Including a Novel Approach to the Design of Combinatorial Libraries Containing Privileged Substructures. *Journal of Medicinal Chemistry* **42**:3251–3264.

Matter H and T Pötter (1999). Comparing 3D Pharmacophore Triplets and 2D Fingerprints for Selecting Diverse Compound Subsets. *Journal of Chemical Information and Computer Sciences* **39**:1211–1225.

McFarland J W and D J Gains (1990). Linear Discriminant Analysis and Cluster Significance Analysis. In Ramsden C A (Editor), *Comprehensive Medicinal Chemistry*, Volume 4. New York, Pergamon Press, pp 667–689.

McKenna J M, F Halley, J E Souness, I M McLay, S D Pickett, A J Collis, K Page and I Ahmed (2002). An Algorithm-Directed Two-Component Library Synthesized via Solid-Phase Methodology Yielding Potent and Orally Bioavailable p38 MAP Kinase Inhibitors. *Journal of Medicinal Chemistry* **45**:2173–2184.

McLay I M, F Halley, J E Souness, J McKenna, V Benning, M Birrell, B Burton, M Belvisi, A Collis, A Constan, M Foster, D Hele, Z Jayyosi, M Kelley, C Maslen, G Miller, M-C Ouldelhkim, K Page, S Phipps, K Pollock, B Porter, A J Ratcliffe, E J Redford, S Webber, B Slater, V Thybaud and N Wilsher (2001). The Discovery of RPR 200765A, a p38 MAP Kinase Inhibitor Displaying a Good Oral Anti-Arthritic Efficacy. *Bioorganic and Medicinal Chemistry* **9**:537–554.

McMahon A J and P M King (1997). Optimization of Carbó Molecular Similarity Index Using Gradient Methods. *Journal of Computational Chemistry* **18**:151–158.

MDDR. MDL Drug Data Report. MDL Information Systems, Inc. 14600 Catalina Street, San Leandro, CA 94577. Also at http://www.mdli.com.

MDL. MDL Information Systems, Inc. 14600 Catalina Street, San Leandro, CA 94577. Also at http://www.mdli.com.

Meehan P and H Schofield (2001). CrossFire: A Structural Revolution for Chemists. *Online Information Review* **25**:241–249.

Meng E C, B K Shoichet and I D Kuntz (1992). Automated Docking with Grid-Based Energy Evaluation. *Journal of Computational Chemistry* **13**:505–524.

Milligan G W (1980). An Examination of the Effect of Six Types of Error in Perturbation on Fifteen Clustering Algorithms. *Psychometrika* **45**:325–342.

Mitchell J B O, R A Laskowski, A Alex, M J Forster and J M Thornton (1999). BLEEP – Potential of Mean Force Describing Protein–Ligand Interactions: II. Calculation of Binding Energies and Comparison with Experimental Data. *Journal of Computational Chemistry* **20**:1177–1185.

Molconn-Z. eduSoft, LC, PO Box 1811, Ashland, VA 23005. Also at http://www.eslc.vabiotech.com.

Moon J B and W J Howe (1990). 3D Database Searching and De Novo Construction Methods in Molecular Design. *Tetrahedron Computer Methodology* **3**:697–711.

Morgan H L (1965). The Generation of a Unique Machine Description for Chemical Structures – A Technique Developed at Chemical Abstracts Service. *Journal of Chemical Documentation* **5**:107–113.

Motoc I, R A Dammkoehler, D Mayer and J Labanowski (1986). Three-Dimensional Quantitative Structure–Activity Relationships, I. General Approach to Pharmacophore Model Validation. *Quantitative Structure–Activity Relationships* **5**:99–105.

Mount J, J Ruppert, W Welch and A N Jain (1999). IcePick: A Flexible Surface-Based System for Molecular Diversity. *Journal of Medicinal Chemistry* **42**:60–66.

Muegge I (2000). A Knowledge-Based Scoring Function for Protein–Ligand Interactions: Probing the Reference State. *Perspectives in Drug Discovery and Design* **20**:99–114.

Muegge I and Y C Martin (1999). A General and Fast Scoring Function for Protein–Ligand Interactions: A Simplified Potential Approach. *Journal of Medicinal Chemistry* **42**:791–804.

Murray C W, T R Auton and M D Eldridge (1998). Empirical Scoring Functions. II. The Testing of an Empirical Scoring Function for the Prediction of Ligand–Receptor Binding Affinities and the Use of a Bayesian Regression to Improve the Quality of the Model. *Journal of Computer-Aided Molecular Design* **12**:503–519.

Murray-Rust P and H Rzepa (1999). Chemical Markup, XML, and the Worldwide Web. 1. Basic Principles. *Journal of Chemical Information and Computer Sciences* **39**:923–942.

Murray-Rust P and J P Glusker (1984). Directional Hydrogen Bonding to sp^2 and sp^3-Hybridized Oxygen Atoms and Its Relevance to Ligand–Macromolecule Interactions. *The Journal of the American Chemical Society* **106**:1018–1025.

Murtaugh F (1983). A Survey of Recent Advances in Hierarchical Clustering Algorithms. *Computer Journal* **26**:354–359.

Murzin A G, S E Brenner, T Hubbard and C Chothia (1995). SCOP: A Structural Classification of Proteins Database for the Investigation of Sequences and Structures. *Journal of Molecular Biology* **247**:536–540.

Nagy M Z, S Kozics, T Veszpremi and P Bruck (1988). Substructure Search on Very Large Files Using Tree-Structured Databases. In Warr W A (Editor), *Chemical Structures. The International Language of Chemistry*. Berlin, Springer, pp 127–130.

Namasivayam S and P M Dean (1986). Statistical Method for Surface Pattern Matching Between Dissimilar Molecules: Electrostatic Potentials and Accessible Surfaces. *Journal of Molecular Graphics* **4**:46–50.

NCI. National Cancer Institute, Suite 3036A, 6116 Executive Boulevard, MSC8322 Bethesda, MD 20892-8322. Also at http// www.nci.nih.gov.

Nicolaou C, S Tamura, B P Kelley, S Bassett and R F Nutt (2002). Analysis of Large Screening Datasets via Aaptively Grown Phylogenic-like Trees. *Journal of Chemical Information and Computer Sciences* **42**:1069–1079.

Nilakantan R, F Immermann and K A Haraki (2002). A Novel Approach To Combinatorial Library Design. *Combinatorial Chemistry and High Throughput Screening* **5**:105–110.

Nilakantan R, N Bauman and R A Venkataraghavan (1993). A New Method for Rapid Characterisation of Molecular Shape: Applications in Drug Design. *Journal of Chemical Information and Computer Sciences* **33**:79–85.

Nilakantan R, N Bauman, J S Dixon and R Ventakaraghavan (1987). Topological Torsion: A New Molecular Descriptor for SAR Applications. Comparison with Other Descriptors. *Journal of Chemical Information and Computer Sciences* **27**:82–85.

Nissink J W M, C Murray, M Hartshorn, M L Verdonk, J C Cole and R Taylor (2002). A New Test Set for Validating Predictions of Protein–Ligand Interaction. *Proteins: Structure Function and Genetics* **49**:457–471.

Norinder U and M Haeberlein (2002). Computational Approaches to the Prediction of the Blood-Brain Distribution. *Advanced Drug Delivery Reviews* **54**:291–313.

Oprea T I (2000). Property Distributions of Drug-Related Chemical Databases. *Journal of Computer-Aided Molecular Design* **14**:251–264.

Oprea T I (Editor) (2005). *Chemoinformatics in Drug Discovery*. Weinheim, Wiley-VCH.

Orengo C A, T P Flores, W R Taylor and J M Thornton (1993). Identification and Classification of Protein Fold Families. *Protein Engineering* **6**:485–500.

Oshiro C M, I D Kuntz and J S Dixon (1995). Flexible Ligand Docking Using a Genetic Algorithm. *Journal of Computer-Aided Molecular Design* **9**:113–130.

Palm K, K Luthman, A-L Ungell, G Strandlund and P Artursson (1996). Correlation of Drug Absorption with Molecular Surface Properties. *Journal of Pharmaceutical Sciences* **85**:32–39.

Paris G (2000). Quoted in Warr W, *Balancing the Needs of the Recruiters and the Aims of the Educators*. Also at http://www.warr.com/wzarc00.html.

Parkar F A and D Parkin (1999). Comparison of Beilstein CrossFire*Plus*Reactions and the Selective Reaction Databases under ISIS. *Journal of Chemical Information and Computer Sciences* **39**:281–288.

Parretti M F, R T Kroemer, J H Rothman and W G Richards (1997). Alignment of Molecules by the Monte Carlo Optimization of Molecular Similarity Indices. *Journal of Computational Chemistry* **18**:1344–1353.

Pastor M, G Cruciani and S Clementi (1997). Smart Region Definition: A New Way to Improve the Predictive Ability and Interpretability of Three-Dimensional Quantitative Structure–Activity Relationships. *Journal of Medicinal Chemistry* **40**:1455–1464.

Patani G A and E J LaVoie (1996). Bioisosterism: A Rational Approach in Drug Design. *Chemical Reviews* **96**:3147–3176.

Patel Y, V J Gillet, G Bravi and A R Leach (2002). A Comparison of the Pharmacophore Identification Programs Catalyst, DISCO and GASP. *Journal of Computer-Aided Molecular Design* **16**:693–681.

Patterson D E, R D Cramer, A M Ferguson, R D Clark and L E Weinberger (1996). Neighbourhood Behaviour: A Useful Concept for Validation of "Molecular Diversity" Descriptors. *Journal of Medicinal Chemistry* **39**:3049–3059.

pdf. Adobe Systems Incorporated, 345 Park Avenue, San Jose, CA 95110–2704. Also at http:/www.adobe.com.

Pearlman R S and K M Smith (1998). Novel Software Tools for Chemical Diversity. *Perspectives in Drug Discovery and Design* **9–11**:339–353.

Pearlman R S and K M Smith (1999). Metric Validation and the Receptor-Relevant Subspace Concept. *Journal of Chemical Information and Computer Sciences* **39**:28–35.

Pepperrell C A and P Willett (1991). Techniques for the Calculation of Three-Dimensional Structural Similarity Using Inter-Atomic Distances. *Journal of Computer-Aided Molecular Design* **5**:455–474.

Pepperrell C A, P Willett and R Taylor (1990). Implementation and Use of an Atom-Mapping Procedure for Similarity Searching in Databases of 3D Chemical Structures. *Tetrahedron Computer Methodology* **3**:575–593.

Perola E and P S Charifson (2004). Conformational Analysis of Drug-Like Molecules Bound to Proteins: An Extensive Study of Ligand Reorganization upon Binding. *Journal of Medicinal Chemistry* **47**:2499–2510.

Perry N C and V J van Geerestein (1992). Database Searching on the Basis of Three-Dimensional Molecular Similarity Using the SPERM Program. *Journal of Chemical Information and Computer Sciences* **32**:607–616.

Pickett S D, I M McLay and D E Clark (2000). Enhancing the Hit-to-Lead Properties of Lead Optimization Libraries. *Journal of Chemical Information and Computer Sciences* **40**:263–272.

Pickett S D, J S Mason and I M McLay (1996). Diversity Profiling and Design Using 3D Pharmacophores: Pharmacophore-Derived Queries (PDQ). *Journal of Chemical Information and Computer Sciences* **36**:1214–1223.

Platts J A, D Butina, M H Abraham and A Hersey (1999). Estimation of Molecular Linear Free Energy Relation Descriptors Using a Group Contribution Approach. *Journal of Chemical Information and Computer Sciences* **39**:835–845.

Poirrette A R, P Willett and F H Allan (1991). Pharmacophoric Pattern Matching in Files of Three-Dimensional Chemical Structures: Characterisation and Use of Generalised Valence Angle Screens. *Journal of Molecular Graphics* **9**:203–217.

Poirrette A R, P Willett and F H Allen (1993). Pharmacophoric Pattern Matching in Files of Three-Dimensional Chemical Structures: Characterisation and Use of Generalised Torsion Angle Screens. *Journal of Molecular Graphics* **11**:2–14.

Poso A, R Juvonen and J Gynther (1995). Comparative Molecular Field Analysis of Compounds with CYP2A5 Binding Affinity. *Quantitative Structure–Activity Relationships* **14**:507–511.

Pötter T and H Matter (1998). Random or Rational Design? Evaluation of Diverse Compound Subsets from Chemical Structure Databases *Journal of Medicinal Chemistry* **41**:478–488.

Press W H, B P Flannery, S A Teukolsky and W T Vetterling (1993). *Numerical Recipes in C: The Art of Scientific Computing*. Cambridge, Cambridge University Press.

Quinlan J R (1993). *C*4.5. San Mateo, CA, Morgan Kaufmann.

Quinlan J R (1996). Bagging, Boosting and C4.5. In *Proceedings of the 13th American Association for Artificial Intelligence National Conference on Artificial Intelligence*. Menlo Park, AAAI Press.

Randić M (1975). On the Characterization of Molecular Branching. *The Journal of the American Chemical Society* **97**:6609–6615.

Randić M (2001). The Connectivity Index 25 Years After. *Journal of Molecular Graphics and Modelling* **20**:19–35.

Randić M and M Razinger (1995). Molecular Topographic Indices. *Journal of Chemical Information and Computer Sciences* **35**:140–147.

Rarey M and J S Dixon (1998). Feature Trees: A New Molecular Similarity Measure Based on Tree Matching. *Journal of Computer-Aided Molecular Design* **12**:471–490.

Rarey M and M Stahl (2001). Similarity Searching in Large Combinatorial Chemistry Spaces. *Journal of Computer-Aided Molecular Design* **15**:497–520.

Rarey M, B Kramer, T Lengauer and G Klebe (1996). A Fast Flexible Docking Method Using an Incremental Construction Algorithm. *Journal of Molecular Biology* **261**:470–489.

Ray L C and R A Kirsch (1957). Finding Chemical Records by Digital Computers. *Science* **126**:814–819.

Raymond J W and P Willett (2002a). Maximum Common Subgraph Isomorphism Algorithms for the Matching of Chemical Structures. *Journal of Computer-Aided Molecular Design* **16**:521–533.

Raymond J W and P Willett (2002b). Effectiveness of Graph-Based and Fingerprint-Based Similarity Measures for Virtual Screening of 2D Chemical Structure Databases. *Journal of Computer-Aided Molecular Design* **16**:59–71.

Raymond J W, E J Gardiner and P Willett (2002). RASCAL: Calculation of Graph Similarity Using Maximum Common Edge Subgraphs. *Computer Journal* **45**:631–644.

Read R C and D G Corneil (1977). The Graph Isomorphism Disease. *Journal of Graph Theory* **1**:339–363.

Rekker R (1977). *The Hydrophobic Fragmental Constant*. Amsterdam, Elsevier Scientific.

Rekker R (1992). *Calculation of Drug Lipophilicity*. Weinheim, VCH.

Richards W G (2002). Innovation: Virtual Screening Using Grid Computing: the Screensaver Project. *Nature Reviews Drug Discovery* **1**:551–555.

Richmond N J, P Willett and R D Clark (2004). Alignment of Three-Dimensional Molecules Using an Image Recognition Algorithm. *Journal of Molecular Graphics and Modelling* **23**:199–209.

Roberts G, G J Myatt, W P Johnson, K P Cross and P E Blower Jr (2000). LeadScope: Software for Exploring Large Sets of Screening Data. *Journal of Chemical Information and Computer Sciences* **40**:1302–1314.

Roche O, P Schneider, J Zuegge, W Guba, M Kansy, A Alanine, K Bleicher, F Danel, E-M Gutknecht, M Rogers-Evans, W Neidhart, H Stalder, M Dillon, E Sjögren, N Fotouhi, P Gillespie, R Goodnow, W Harris, P Jones, M Taniguchi, S Tsujii, W von der Saal, G Zimmermann and G Schneider (2002). Development of a Virtual Screening Method for Identification of "Frequent Hitters" in Compound Libraries. *Journal of Medicinal Chemistry* **45**:137–142.

Rogers D, R D Brown and M Hahn (2005). Using Extended-Connectivity Fingerprints with Laplacian-Modified Bayesian Analysis in High-Throughput Screening Follow-Up. *Journal of Biomolecular Screening* **10**:682–686.

Rogers D and A J Hopfinger (1994). Application of Genetic Function Approximation to Quantitative Structure–Activity Relationships and Quantitative Structure–Property Relationships. *Journal of Chemical Information and Computer Sciences* **34**:854–866.

Rosenkranz H S, A R Cunningham, Y P Zhang, H G Claycamp, O T Macina, N B Sussman, S G Grant and G Klopman (1999). Development, Characterization and Application of Predictive-Toxicology Models. *SAR and QSAR* **10**:277–298.

Rubin V and P Willett (1983). A Comparison of Some Hierarchical Monothetic Divisive Clustering Algorithms for Structure–Property Correlation. *Analytica Chemica Acta* **151**:161–166.

Rummelhart D E, G W Hinton and R J Williams (1986). Learning Representations by Back-Propagating Errors. *Nature* **323**:533–536.

Rush T S, J A Grant, L Mosyak and A Nicholls (2005). A Shape-Based 3-D Scaffold Hopping Method and Its Application to a Bacterial Protein–Protein Interaction. *Journal of Medicinal Chemistry* **48**:1489–1495.

Rusinko A III, J M Skell, R Balducci, C M McGarity and R S Pearlman (1988). CONCORD: A Program for the Rapid Generation of High Quality 3D Molecular Structures. St Louis, MO, The University of Texas at Austin and Tripos Associates.

Rusinko A III, M W Farmen, C G Lambert, P L Brown and S S Young (1999). Analysis of a Large Structure/Biological Activity Data Set Using Recursive Partitioning. *Journal of Chemical Information and Computer Sciences* **39**:1017–1026.

Russo E (2002). Chemistry Plans a Structural Overhaul. *Nature* **419** (12 September2002) Naturejobs:4–7.

Sadowski J and H Kubinyi (1998). A Scoring Scheme for Discriminating Between Drugs and Nondrugs. *Journal of Medicinal Chemistry* **41**:3325–3329.

Saeh J C, P D Lyne, B K Takasaki and D A Cosgrove (2005). Lead Hopping Using SVM and 3d Pharmacophore Fingerprints. *Journal of Chemical Information and Modeling* **45**:1122–1133.

Sammon J W Jr (1969). A Nonlinear Mapping for Data Structure Analysis. *IEEE Transactions on Computers* **18**:401–409.

Sanderson D M and C G Earnshaw (1991). Computer Prediction of Possible Toxic Action form Chemical Structure; The DEREK System. *Human and Experimental Toxicology* **10**:261–273.

Sanz F, F Manaut, J Rodriguez, E Loyoza and E Ploez-de-Brinao (1993). MEPSIM: A Computational Package for Analysis and Comparison of Molecular Electrostatic Potentials. *Journal of Computer-Aided Molecular Design* **7**:337–347.

Saunders M (1987). Stochastic Exploration of Molecular Mechanics Energy Surface: Hunting for the Global Minimum. *The Journal of the American Chemical Society* **109**:3150–3152.

Saunders M, K N Houk, Y-D Wu, W C Still, M Lipton, G Chang and W C Guida (1990). Conformations of Cycloheptadecane. A Comparison of Methods for Conformational Searching. *The Journal of the American Chemical Society* **112**:1419–1427.

Scheraga H A (1993). Searching Conformational Space. In van Gunsteren W F, P K Weiner and A J Wilkinson (Editors), *Computer Simulation of Biomolecular Systems*, Volume 2. Leiden, ESCOM Science.

Schneider G (2000). Neural Networks are Useful Tools for Drug Design. *Neural Networks* **13**:15–16.

Schneider G and P Wrede (1998). Artificial Neural Networks for Computer-Based Molecular Design. *Progress in Biophysics and Molecular Biology* **70**:175–222.

Schneider G, W Neidhart, T Giller and G Schmid (1999). "Scaffold-Hopping" by Topological Pharmacophore Search: A Contribution to Virtual Screening. *Angewandte Chemie (International ed. in English)* **38**:2894–2896.

Schofield H, G Wiggins and P Willett (2001). Recent Developments in Chemoinformatics Education. *Drug Discovery Today* **6**:931–934.

Schuffenhauer A, V J Gillet and P Willett (2000). Similarity Searching in Files of Three-Dimensional Chemical Structures: Analysis of the BIOSTER Database Using Two-Dimensional Fingerprints and Molecular Field Descriptors. *Journal of Chemical Information and Computer Sciences* **40**:295–307.

Schuffenhauer A, P Floersheim, P Acklin and E Jacoby (2003). Similarity Metrics for Ligands Reflecting the Similarity of the Target Proteins. *Journal of Chemical Information and Computer Sciences* **43**:391–405.

SciTegic, 9665 Chesapeake Drive, Suite 401, San Diego, CA 92123-1365. Also at http://www.scitegic.com.

Shemetulskis N, D Weininger, C J Blankley, J J Yang and C Humblet (1996). Stigmata: An Algorithm to Determine Structural Commonalities in Diverse Datasets. *Journal of Chemical Information and Computer Sciences* **36**:862–871.

Shenkin P S, D L Yarmusch, R M Fine, H Wang and C Levinthal (1987). Predicting Antibody Hypervariable Loop Conformation. I. Ensembles of Random Conformations for Ringlink Structures. *Biopolymers* **26**:2053–2085.

Sheridan R P and S K Kearsley (1995). Using a Genetic Algorithm to Suggest Combinatorial Libraries. *Journal of Chemical Information and Computer Sciences* **35**:310–320.

Sheridan R P and S K Kearsley (2002). Why Do We Need So Many Chemical Similarity Search Methods? *Drug Discovery Today* **7**:903–911.

Sheridan R P, B P Feuston, V N Maiorov and S K Kearsley (2004). Similarity to Molecules in the Training Set is a Good Discriminator for Prediction Accuracy in QSAR. *Journal of Chemical Information and Computer Sciences* **44**:1912–1928.

Sheridan R P, M D Miller, D J Underwood and S K Kearsley (1996). Chemical Similarity Using Geometric Atom Pair Descriptors. *Journal of Chemical Information and Computer Sciences* **36**:128–136.

Sheridan R P, S G SanFeliciano and S K Kearsley (2000). Designing Targeted Libraries with Genetic Algorithms. *Journal of Molecular Graphics and Modelling* **18**:320–334.

Shi L M, Y Fan, J K Lee, M Waltham, D T Andrews, U Scherf, K D Paull and J N Weinstein (2000). Mining and Visualizing Large Anticancer Drug Discovery Databases. *Journal of Chemical Information and Computer Sciences* **40**:367–379.

Shoichet B K, S L McGovern, B Wei and J J Irwin (2002). Lead Discovery Using Molecular Docking. *Current Opinion in Chemical Biology* **6**:439–446.

Silipo C and Vittoria A (1990). Three-Dimensional Structure of Drugs. In Ramsden C A (Editor), *Comprehensive Medicinal Chemistry*, Volume 4. New York, Pergamon Press, pp 153–204.

Smellie A S, S D Kahn and S L Teig (1995a). Analysis of Conformational Coverage. 1. Validation and Estimation of Coverage. *Journal of Chemical Information and Computer Sciences* **35**:285–294.

Smellie A S, S L Teig and P Towbin (1995b). Poling: Promoting Conformational Variation. *Journal of Computational Chemistry* **16**:171–187.

Snarey M, N K Terrett, P Willett and D J Wilton (1997). Comparison of Algorithms for Dissimilarity-Based Compound Selection. *Journal of Molecular Graphics and Modelling* **15**:372–385.

Sneath P H A and R R Sokal (1973). *Numerical Taxonomy*. San Francisco, CA, W H Freeman.

So S S, S P van Helden, V J van Geerstein and M Karplus (2000). Quantitative Structure–Activity Relationship Studies of Progesterone Receptor Binding Steroids. *Journal of Chemical Information and Computer Sciences* **40**:762–772.

Sotriffer C A, H Gohlke and G Klebe (2002). Docking into Knowledge-Based Potential Fields: A Comparative Evaluation of DrugScore. *Journal of Medicinal Chemistry* **45**:1967–1970.

Spotfire. Spotfire, 212 Elm Street, Somerville, MA 02144. Also at http://www.spotfire.com.

SPRESI. The Spresi database. Daylight Chemical Information Systems, Inc., Mission Viejo, CA. Also at http://www.daylight.com.

Srinivasan J, A Castellino, E K Bradley, J E Eksterowicz, P D J Grootenhuis, S Putta and R V Stanton (2002). Evaluation of a Novel Shape-Based Computational Filter for Lead Evolution: Application to Thrombin Inhibitors. *Journal of Medicinal Chemistry* **45**:2494–2500.

Stahl M (2000). Structure-Based Library Design. *Methods and Principles in Medicinal Chemistry* **10**:229–264.

Stahl M and H-J Böhm (1998). Development of Filter Functions for Protein–Ligand Docking. *Journal of Molecular Graphics and Modelling* **16**:121–132.

Stahl M and M Rarey (2001). Detailed Analysis of Scoring Functions for Virtual Screening. *Journal of Medicinal Chemistry* **44**:1036–1042.

Stein S E, S R Heller and D Tchekhovskoi (2003). An Open Standard for Chemical Structure Representation: The IUPAC Chemical Identifier. In *Proceedings of the 2003 International Chemical Information Conference*. Nimes, France, October 19–22; Infonortics: Tetbury, UK, pp 131–143.

Stouch, R R, J R Kenyon, S R Johynson, X-Q Chen, A Doweyko and Y Li (2003). *In Silico* ADME/Tox: Why Models Fail. *Journal of Computer-Aided Molecular Design* **17**:83–92.

Stouch T R and A Gudmundsson (2002). Progress in Understanding the Structure–Activity Relationships of P-glycoprotein. *Advanced Drug Delivery Reviews* **54**:315–328.

Swain C G and E C Lupton (1968). Field and Resonance Components of Substituent Effects. *The Journal of the American Chemical Society* **90**:4328–4337.

Swain C G, S H Unger, N R Rosenquist and M S Swain (1983). Substituent Effects on Chemical Reactivity. Improved Evaluation of Field and Resonance Components. *The Journal of the American Chemical Society* **105**:492–502.

Takahashi Y, M Sukekawa and S Sasaki (1992). Automatic Identification of Molecular Similarity Using Reduced-Graph Representation of Chemical Structure. *Journal of Chemical Information and Computer Sciences* **32**:639–643.

Tamura S Y, P A Bacha, H S Gruver and R F Nutt (2002). Data Analysis of High-Throughput Screening Results: Application of Multidomain Clustering to the NCI Anti-HIV Data Set. *Journal of Medicinal Chemistry* **45**:3082–3093.

Taylor R (1995). Simulation Analysis of Experimental Design Strategies for Screening Random Compounds as Potential New Drugs and Agrochemicals. *Journal of Chemical Information and Computer Sciences* **35**:59–67.

Taylor R D, P J Jewsbury and J W Essex (2002). A Review of Protein-Small Molecule Docking Methods. *Journal of Computer-Aided Molecular Design* **16**:151–166.

Teague S J, A M Davis, P D Leeson and T Oprea (1999). The Design of Leadlike Combinatorial Libraries. *Angewandte Chemie (International ed. in English)* **38**:3743–3748.

ter Laak A, J Venhorst, G M Donne-Op Den Kelder and H Timmerman (1995). The Histamine H1-Receptor Antagonist Binding Site. A Stereoselective Pharmacophoric Model Based upon (Semi-)Rigid H1-Antagonists and Including a Known Interaction Site on the Receptor. *Journal of Medicinal Chemistry* **38**:3351–3360.

Terp G E, B N Johansen, I T Christensen and F S Jorgensen (2001). A New Concept for Multidimensional Selection of Ligand Conformations (MultiSelect) and Multidimensional Scoring (MultiScore) of Protein–Ligand Binding Affinities. *Journal of Medicinal Chemistry* **44**:2333–2343.

Terrett N K (1998). *Combinatorial Chemistry*. Oxford, Oxford University Press.

Thibaut U, G Folkers, G Klebe, H Kubinyi, A Merz and D Rognan (1993). Recommendations for CoMFA Studies and 3D QSAR Publications. In Kubinyi H (Editor), *3D QSAR in Drug Design*. Leiden, ESCOM, pp 711–728.

Thornber C W (1979). Isosterism and Molecular Modification in Drug Design. *Chemical Society Reviews* 8:563–580.

Thorner D A, D J Wild, P Willett and P M Wright (1996). Similarity Searching in Files of Three-dimensional Chemical Structures: Flexible Field-based Searching of Molecular Electrostatic Potentials. *Journal of Chemical Information and Computer Sciences* 36:900–908.

Thorner D A, P Willett, P M Wright and R Taylor (1997). Similarity Searching in Files of Three-Dimensional Chemical Structures: Representation and Searching of Molecular Electrostatic Potentials Using Field-Graphs. *Journal of Computer-Aided Molecular Design* 11:163–174.

Tice C M (2001). Selecting the Right Compounds for Screening: Does Lipinsk's Rule of 5 for Pharmaceuticals Apply to Agrochemicals? *Pest Management Science* 57:3–16.

Todeschini R, V Consonni and M Pavan (2004). A Distance Measure Between Models: A Tool for Similarity/Diversity Analysis of Model Populations. *Chemometrics and Intelligent Laboratory Systems* 70:55–61.

Tong W, D R Lowis, R Perkins, Y Chen, W J Welsh, D W Goddette, T W Heritage and D M Sheehan (1998). Evaluation of Quantitative Structure–Activity Relationship Methods for Large-Scale Prediction of Chemicals Binding to the Estrogen Receptor. *Journal of Chemical Information and Computer Sciences* 38:669–688.

Triballeau N, J Acher, I Brabet, J-P Pin and H-O Bertrand (2005). Virtual Screening Workflow Development Guided by the "Receiver Operating Characteristic" Curve Approach. Application to High-Throughput Docking on Metabotropic Glutamate Receptor Subtype 4. Journal of Medicinal Chemistry 48:2534–2547.

Trinajstic N (Editor) (1983). *Chemical Graph Theory*. Boca Raton, FL, CRC Press.

Tute M S (1990). History and Objectives of Quantitative Drug Design. In Ramsden C A (Editor), *Comprehensive Medicinal Chemistry*, Volume 4. New York, Pergamon Press, pp 1–31.

Tversky A (1977). Features of Similarity. *Psychological Reviews* 84:327–352.

Ullmann J R (1976). An Algorithm for Subgraph Isomorphism. *Journal of the Association for Computing Machinery* 23:31–42.

UNITY. Tripos Inc., 1699 South Hanley Road, St. Louis, MO 63144-2913. Also at http://www.tripos.com.

Valler M J and D V S Green (2000). Diversity Screening Versus Focussed Screening in Drug Discovery. *Drug Discovery Today* 5:286–293.

van de Waterbeemd (2002). High-Throughput and *In Silico* Techniques in Drug Metabolism and Pharmacokinetics. *Current Opinion in Drug Discovery and Development* 5:33–43.

van de Waterbeemd H, D A Smith, K Beaumont and D K Walker (2001). Property-Based Design: Optimisation of Drug Absorption and Pharmacokinetics. *Journal of Medicinal Chemistry* 44:1313–1333.

van de Waterbeemd H, G Camenisch, G Folkers and O A Raevsky (1996). Estimation of Caco-2 Cell Permeability using Calculated Molecular Descriptors. *Quantitative Structure–Activity Relationships* 15:480–490.

van Drie J H (2003). Pharmacophore Discovery – Lessons Learned. *Current Pharmaceutical Design* 9:1649–1664.

van Geerestein V J, N C Perry, P G Grootenhuis and C A G Haasnoot (1990). 3D Database Searching on the Basis of Ligand Shape Using the SPERM Prototype Method. *Tetrahedron Computer Methodology* 3:595–613.

van Rhee A M, J Stocker, D Printzenhoff, C Creech, P K Wagoner and K L Spear (2001). Retrospective Analysis of an Experimental High-Throughput Screening Data Set by Recursive Partitioning. *Journal of Combinatorial Chemistry* **3**:267–277.

Veber D F, S R Johnson, H-Y Cheng, B R Smith, K W Ward and K D Kopple (2002). Molecular Properties that Influence the Oral Bioavailability of Drug Candidates. *Journal of Medicinal Chemistry* **45**:2615–2623.

Wagener M and V J van Geerestein (2000). Potential Drugs and Nondrugs: Prediction and Identification of Important Structural Features. *Journal of Chemical Information and Computer Sciences* **40**:280–292.

Waldman M, H Li and M Hassan (2000). Novel Algorithms for the Optimization of Molecular Diversity of Combinatorial Libraries. *Journal of Molecular Graphics and Modelling* **18**:412–426.

Walters W P and M A Murcko (2002). Prediction of "Drug-Likeness". *Advanced Drug Delivery Reviews* **54**:255–271.

Wang R, Y Fu and L Lai (1997). A New Atom-Additive Method for Calculating Partition Coefficients. *Journal of Chemical Information and Computer Sciences* **37**:615–621.

Ward J H (1963). Hierarchical Grouping to Optimise an Objective Function. *The Journal of the American Statistical Association* **58**:236–244.

Warmuth M K, J Liao, G Ratsch, M Mathieson, S Putta and C Lemmen (2003). Active Learning with Support Vector Machines in the Drug Discovery Process. *Journal of Chemical Information and Computer Sciences* **43**:667–673.

Warren G L, C W Andrews, A-M Capelli, B Clarke, J LaLonde, M H Lambert, M Lindvall, N Nevins, S F Semus, S Senger, G Tedesco, I D Wall, J M Woolven, C E Peishoff and M S Head (2006). A Critical Assessment of Docking Programs and Scoring Functions. *Journal of Medicinal Chemistry* **49**:5912–5931.

WDI. World Drug Index, Thomson Derwent. 14 Great Queen Street, London W2 5DF, UK. Also at http://www.derwent.com.

Weininger D (1988). SMILES, a Chemical Language and Information System. 1. Introduction to Methodology and Encoding Rules. *Journal of Chemical Information and Computer Sciences* **28**:31–36.

Weininger D, A Weininger and J L Weininger (1989). SMILES. 2. Algorithm for Generation of Unique SMILES Notation. *Journal of Chemical Information and Computer Sciences* **29**:97–101.

Welch W, J Ruppert and A N Jain (1996). Hammerhead: Fast, Fully Automated Docking of Flexible Ligands to Protein Binding Sites. *Chemistry and Biology* **3**:449–462.

Whittle M, V J Gillet, P Willett, A Alex and J Loesel (2004). Enhancing the Effectiveness of Virtual Screening by Fusing Nearest Neighbor Lists: A Comparison of Similarity Coefficients. *Journal of Chemical Information and Computer Sciences* **44**:1840–1848.

Wiener H. (1947). Structural Determination of Paraffin Boiling Point. *The Journal of the American Chemical Society* **69**:17–20.

Wild D J and C J Blankley (2000). Comparison of 2D Fingerprint Types and Hierarchy Level Selection Methods for Structural Grouping Using Ward's Clustering. *Journal of Chemical Information and Computer Sciences* **40**:155–162.

Wild D J and P Willett (1996). Similarity Searching in Files of Three-Dimensional Chemical Structures: Alignment of Molecular Electrostatic Potentials with a Genetic Algorithm. *Journal of Chemical Information and Computer Sciences* **36**:159–167.

Wildman S A and G M Crippen (1999). Prediction of Physicochemical Parameters by Atomic Contributions. *Journal of Chemical Information and Computer Sciences* **39**:868–873.

Willett P (1979). A Screen Set Generation Algorithm. *Journal of Chemical Information and Computer Sciences* **19**:159–162.

Willett P (1987). *Similarity and Clustering in Chemical Information Systems.* Letchworth, Research Studies Press.

Willett P (1999). Subset Selection Methods for Chemical Databases. In Dean P M and R A Lewis (Editors), *Molecular Diversity in Drug Design.* Dordrecht, Kluwer Academic, pp 115–140.

Willett P (2006). Enhancing the Effectiveness of Ligand-Based Virtual Screening Using Data Fusion. *QSAR & Combinatorial Science* **25**:1143–1152.

Willett P, J M Barnard and G M Downs (1998). Chemical Similarity Searching. *Journal of Chemical Information and Computer Sciences* **38**:983–996.

Willett P, V Winterman and D Bawden (1986). Implementation of Nearest Neighbour Searching in an Online Chemical Structure Search System. *Journal of Chemical Information and Computer Sciences* **26**:36–41.

Wilton D, P Willett, K Lawson and G Mullier (2003). Comparison of Ranking Methods for Virtual Screening in Lead-Discovery Programmes. *Journal of Chemical Information and Computer Sciences* **43**:469–474.

Wipke W T and T M Dyott (1974). Stereochemically Unique Naming Algorithm. *The Journal of the American Chemical Society* **96**:4825–4834.

Wipke W T, S Krishnan and G I Ouchi (1978). Hash Functions for Rapid Storage and Retrieval of Chemical Structures. *Journal of Chemical Information and Computer Sciences* **18**:32–37.

Wiswesser W J (1954). *A Line-Formula Chemical Notation,* New York, Crowell.

Wold H (1982). Soft Modeling. The Basic Design and Some Extensions. In Joreskog K-G and H Wold (Editors), *Systems Under Indirect Observation.* Amsterdam, North-Holland.

Wold S (1994). PLS for Multivariate Linear Modeling. In van de Waterbeemd H (Editor), *QSAR.* Weinheim, Verlag-Chemie.

Wold S, E Johansson and M Cocchi (1993). PLS – Partial Least-Squares Projections to Latent Structures. In Kubinyi H (Editor), *3D QSAR* Leiden, ESCOM, pp 523–550.

Xia X Y, E G Maliski, P Gallant and D Rogers (2004). Classification of Kinase Inhibitors Using a Bayesian Model. *Journal of Medicinal Chemistry* **47**:4463–4470.

Xiang X-D. X Sun, G Briceno, Y Lou, K-A Wang, H Chang, W G Wallace-Freedman, S-W Chen and P G Schultz (1995). A Combinatorial Approach to Materials Discovery. *Science* **268**:1738–1740.

Xie D, A Tropsha and T Schlick (2000). An Efficient Projection Protocol for Chemical Databases: Singular Value Decomposition Combined with Truncated-Newton Minimisation. *Journal of Chemical Information and Computer Sciences* **40**:167–177.

Xu Y J and H Gao (2003). Dimension Related Distance and its Application in QSAR/QSPR Model Error Estimation. *QSAR and Combinatorial Science* **22**:422–429.

Xu Y J and M Johnson (2002). Using Molecular Equivalence Numbers to Visually Explore Structural Features That Distinguish Chemical Libraries. *Journal of Chemical Information and Computer Sciences* **42**:912–926.

Xue L, J Godden and J Bajorath (1999). Database Searching for Compounds with Similar Biological Activity Using Short Binary Bit String Representations of Molecules. *Journal of Chemical Information and Computer Sciences* **39**:881–886.

Yutani K, K Ogasahara, T Tsujita and Y Sugino (1987). Dependence of Conformational Stability on Hydrophobicity of the Amino Acid Residue in a Series of Variant Proteins Substituted at a Unique Position of Tryptophan Synthase α Subunit. *Proceedings of the National Academy of Sciences of the United States of America* **84**:4441–4444.

Zass E (1990). A User's View of Chemical Reaction Information Sources. *Journal of Chemical Information and Computer Sciences* **30**:360–372.

Zhang J-H, T D Y Chung and K R Oldenburg (1999). A Simple Statistical Parameter for Use in Evaluation and Validation of High Throughput Screening Assays. *Journal of Biomolecular Screening* 4:67–73.

Zheng W, S J Cho and A Tropsha (1998). Rational Combinatorial Library Design. 1. Focus-2D: A New Approach to the Design of Targeted Combinatorial Chemical Libraries. *Journal of Chemical Information and Computer Sciences* 38:251–258.

Zheng W, S T Hung, J T Saunders and G.L. Seibel (2000). PICCOLO: A Tool for Combinatorial Library Design via Multicriterion Optimization. In Atlman R B, A K Dunkar, L Hunter, K Lauderdale and T E Klein (Editors), *Pacific Symposium on Biocomputing 2000*. Singapore, World Scientific, pp 588–599.

Zupan J and J Gasteiger (1999). *Neural Networks for Chemists*. Weinheim, Wiley-VCH.

SUBJECT INDEX